Molecular Methods for Microbial Identification and Typing

K. J. Towner and A. Cockayne

Department of Microbiology and PHLS Laboratory, University Hospital, Queen's Medical Centre, Nottingham, UK

SPRINGER-SCIENCE+BUSINESS MEDIA, B.V.

First edition 1993

Originally published by Chapman & Hall in 1993
Phototypeset in 10/12pt Palatino by Intype, London

ISBN 978-0-412-49390-4 ISBN 978-94-011-1506-3 (eBook)
DOI 10.1007/978-94-011-1506-3

A catalogue record for this book is available from the British Library

Library of Congress Cataloging-in-Publication data available

∞ Printed on permanent acid-free text paper, manufactured in accordance with the proposed ANSI/NISO Z 39.48–199X and ANSI Z 39.48–1984

Contents

Preface

This book is intended as a source book for postgraduate students, research workers and diagnostic microbiologists in all fields of microbiology, but particularly those active in medical microbiology, who wish to explore the possibilities of modern molecular methods for microbial identification and typing.

Many examples of such applications are concealed in the scientific literature, but rather than dwell in depth on specific organisms or epidemiological situations, we have tried to introduce the principles of each individual method and give an overview of the methodology, advantages and disadvantages implicit in each technique. Thus it is not a 'recipe' book, but is intended as a first point of reference for microbiologists seeking an introduction to these techniques.

The development of these methods has resulted in an exciting period for microbial identification and typing. While many of the methods have been developed with bacteria in mind, we hope that the lists of organisms and extensive references to detailed methodology will enable the interested reader to find an advanced molecular technique that can be applied readily to their own 'favourite' organism, be it a bacterium, virus, fungus or protozoan.

It is a pleasure to acknowledge the encouragement received from friends and colleagues in our own department and elsewhere. We would particularly like to thank Debbie Cooke and Sarinder Day for their help with the illustrations. This book would not have been possible without the tolerance and sustenance provided by Janet Towner and Fiona MacDonald.

Kevin Towner and Alan Cockayne
Nottingham

An introduction to microbial identification and typing methods

1

1.1 A BRIEF OVERVIEW OF MICROBIAL IDENTIFICATION

Early classification and identification methods for microbes relied on tests that were based on the phenotypic properties of individual organisms. Strains were assigned to somewhat intuitive classification groups on the basis of small numbers of characteristics which could be examined easily, such as staining properties, morphology, motility, nutritional requirements, acid production, pigmentation and spore formation. Such an approach suffers from the basic drawback that tests derived for one group of organisms are not always useful for other groups. Thus, for example, many tests used for identifying different members of the Enterobacteriaceae are not suitable for identification of strictly aerobic Gram-negative bacteria. In addition, the introduction of increasingly objective modern taxonomic methods has led to recognition of the fact that over-dependence on a small number of properties for classification purposes can lead to serious errors; indeed, while some taxonomic groupings based on phenotypic properties have stood the test of time, other groups have been shown to contain a wide range of heterogeneous organisms.

Awareness of the limitations of traditional taxonomic methods has resulted in a re-evaluation of the whole process of microbial identification and classification. What began as a rather subjective and intuitive subject has changed radically in recent years. Increasing use of the molecular identification and typing methods which form the main subject of this book has led to revision of some accepted classification schemes and the recognition of new relationships among microbes.

However, whatever the criteria for taxonomic groupings, the basic unit for the purposes of microbial identification remains the **species**.

Traditionally, a microbial species has been considered to be a group of strains which show a high degree of similarity in terms of their phenotypic properties. As indicated above, this is a rather subjective grouping (a **taxospecies**) which may be influenced greatly by the ideas and aims of the taxonomist. Alternative methods of grouping include the **genomic species**, a group of organisms with high DNA–DNA homology values, and the **genospecies**, a group of organisms capable of genetic exchange.

In recent years it has become accepted by microbial taxonomists that nucleic acid hybridization and sequencing studies provide the best available and most rational methods for designating species and determining relationships between different organisms. A complete DNA sequence forms the reference standard to determine phylogeny, and phylogeny should determine taxonomy. A formal molecular definition of a species has been suggested (Wayne *et al.*, 1987), in that a species should, generally, include strains with approximately 70% or greater DNA–DNA relatedness, and with 5°C or less divergence values (ΔT_m); however, only species which can be differentiated by phenotypic properties should be given a formal name. Murray *et al.* (1990) have pointed out that it is completely impracticable to define genera and species solely on the basis of phylogenetic data, and that initial characterization of genera needs to be on the basis of phenotypic properties. Nevertheless, these authors recommended that existing classifications should be re-examined on a molecular basis. The techniques used to determine such molecular relationships are described in detail in later chapters of this book. Obviously, while it is desirable to have as few changes as possible to existing taxonomic arrangements, it is important that molecular relationships should be reflected accurately in nomenclature and classification schemes.

Once individual species have been defined, they can then be arranged into genera and families, on the basis of overall similarity, to form a hierarchical system. However, although identification to the species level is the primary purpose of all microbial classification schemes, the separation and accurate recognition of subtypes within a species is assuming greater importance in all branches of microbiology, and particularly in medical microbiology. Indeed, in many cases, control of communicable diseases would not be possible without the use of typing methods to help define the sources of infection, mechanisms of transmission and rate of spread of infection in a susceptible population. Other areas in which accurate microbial typing is important include ecological studies involving the release and monitoring of novel microorganisms in new natural habitats (Jain *et al.*, 1988), and industrial screening programmes searching for new microbial products (Goodfellow and O'Donnell, 1989).

As indicated earlier, a complete DNA sequence would form the ultimate reference standard for recognizing sub-types within a species. Short of achieving this ideal, any typing technique relies on finding detectable differences between isolates. There is a tendency to assume that organisms which are indistinguishable by a chosen typing method are identical and have a common source. However, it should always be remembered that such organisms may be indistinguishable only when 'viewed' through the typing 'window' generated by a particular typing scheme, and may be totally different when typed by an alternative method. Neither typing system generates 'wrong' results, but each provides a different way or angle of looking at the same problem.

As with identification to the species level, many well-established microbial typing schemes, particularly in clinical microbiology, are based on the recognition of particular phenotypic properties. While this book is concerned primarily with modern molecular methods for microbial identification and typing, it is important that the newer methods should be considered in comparison with the older established methods which they may eventually supersede.

1.2 CONVENTIONAL BIOLOGICAL METHODS FOR MICROBIAL TYPING

Conventional biological typing methods for microorganisms were developed originally in response to a growing awareness of the different factors which influence the spread of human pathogens in various environments. An ideal typing system (Hawkey, 1989) should: (a) be able to type the vast majority of strains encountered; (b) have good discrimination with the ability to recognize a reasonable number of types; (c) show good reproducibility over a long period of time and in different centres; (d) be readily applicable to natural isolates as opposed to laboratory collections of strains; (e) should not be too complicated or expensive. As indicated by the examples in Tables 1.1–1.4, the main biological typing techniques of biotyping, phage typing, serotyping and bacteriocin typing are well established and have been applied to a wide range of microorganisms. Details of the various technical modifications required for each individual genus can be found in the references indicated; however, certain advantages and disadvantages are linked directly to each of the main typing methods, and these are described briefly in the following sections.

1.2.1 BIOTYPING

Frequently, initial differentiation within a newly-delineated species is achieved by examining the cultural and biochemical characteristics of a large collection of individual strains belonging to the species. Such characteristics may include colonial morphology, growth requirements, fermentation ability, carbon source utilization and antibiotic resistances.

In theory such properties are easy to determine, but in practice the determination of biotypes may not always be straightforward. Differences in colonial morphology are often extremely subtle and, therefore, can be rather subjective. Biochemical reactions are more dependable, but may also be difficult to interpret. Many laboratories use commercially available galleries of tests, such as the API 20E system with 20 different tests, to provide a biochemical profile from which different biotypes can be identified. However, apart from instability problems resulting from occasional loss of plasmids encoding certain metabolic functions, variations in the duration of incubation and the inoculum size may affect interpretation of the results, while strains that are freshly isolated may exhibit different reactions compared with strains that have been stored. Newer automated biochemical fingerprinting systems, based on an evaluation of the kinetics of biochemical reactions performed in microtitre plates, have been reported to give good reproducibility and discrimination for certain species, such as *Enterobacter cloacae* (Kühn, Tullis and Burman, 1991), but in many cases there may be only a very limited number of biochemical types within a species, resulting partly from the fact that traditional taxonomic procedures, particularly in clinical microbiology, may have relied on a single specific biochemical reaction to define the species in the first place.

Resistance to a range of antibiotics (the **antibiogram**), while normally easy to determine, is often associated with potentially unstable extra-chromosomal R plasmids, and therefore may be subject to frequent and abrupt multiple changes. The relatively small number of agents with the potential to give different results also means that discrimination is poor. Thus, while it is useful to identify recurring resistance patterns in a local situation, such patterns have little value when considered in isolation in comparative studies involving several different centres.

In general, it can be concluded that the determination of biotypes within a species, while possibly helpful in a limited short-term local investigation, will be useful only in association with other typing methods for long-term investigations covering different epidemiological outbreaks or ecological situations. Examples of organisms in which biotyping schemes have proved useful for epidemiological purposes are listed in Table 1.1.

Table 1.1 Examples of microorganisms in which biotyping schemes have proved useful for epidemiological purposes

Organism	Reference
Acinetobacter spp.	Bouvet (1991)
Aspergillus fumigatus	Fanti *et al.* (1989)
Bacillus spp.	Berkeley *et al.* (1984)
Brucella spp.	McCullough (1978)
Campylobacter spp.	Kapperud *et al.* (1984)
Citrobacter spp.	Lányi (1984)
Neisseria gonorrhoeae	Catlin (1978)
Pasteurella haemolytica	Biberslein (1978)
Salmonella spp.	Katouli *et al.* (1992)
Serratia liquefaciens	Traub (1991)
Streptococcus faecalis	Smyth *et al.* (1987)
Vibrio cholerae	Mukerjee (1978)

1.2.2 PHAGE TYPING

Bacteriophages (phages) are viruses capable of infecting bacteria, leading in some cases to lysis of the bacterial cell and release of further infective phage particles, while in other cases the phage can persist as a relatively stable **prophage** within the bacterial cell. Phage typing is a method for bacterial strain identification that is based upon sensitivity to defined collections of bacteriophages which have been selected to provide the maximum sensitivity for differentiating strains within a particular species. Typing may be 'direct', i.e. based on direct sensitivity to either unadapted or adapted phages, or 'indirect', i.e. based on detection and identification of phages present as prophages in bacteria.

Phage typing schemes are highly sensitive and have been developed for numerous bacterial genera (Table 1.2), many of which have not yet been typed successfully by other means. Such typing schemes are used widely for epidemiological purposes throughout the world. Phage typing remains the major typing method for *Staphylococcus aureus*, and is also particularly useful for subdividing serotypes of *Pseudomonas aeruginosa* and *Salmonella/Shigella* spp. Nevertheless, the methodology has a number of important disadvantages and limitations (Meitert and Meitert, 1978). Since environmental conditions can affect the sensitivity of bacteria to infection by phages, phage typing is a technically complex method in which many variables must be controlled. Some species may contain too few phage types, while others may contain too many phage types represented by only one strain. In other cases, e.g., *Acinetobacter* (Bouvet, 1991), a phagetyping scheme may be reliable with strains from one country, but may be less effective, with a high proportion of non-typable strains, with isolates from other parts of the world.

Table 1.2 Examples of microorganisms for which phage typing schemes have been developed

Organism	Reference
Acinetobacter spp.	Bouvet (1991)
Bacillus spp.	Berkeley *et al.* (1984)
Brucella spp.	Corbel (1984)
Clostridium difficile	Bacon *et al.* (1988)
Corynebacterium diphtheriae	Saragea *et al.* (1979)
Enterobacter cloacae	Verschraegen *et al.* (1988)
Erysipelothrix insidiosa	Stoeve (1979)
Escherichia coli	Milch (1978)
Klebsiella spp.	Slopek (1978)
Listeria monocytogenes	McLauchlin *et al.* (1986)
Mycobacterium spp.	Redmond *et al.* (1979)
Proteus spp.	Bergan (1978b)
Pseudomonas aeruginosa	Bergan (1978a)
Salmonella spp.	Guinée and van Leeuwen (1978)
Shigella spp.	Bergan (1979)
Staphylococcus aureus	Parker (1972)
Streptococcus agalactiae	Stringer (1984)
Streptococcus faecalis	Smyth *et al. (1987)*
Vibrio cholerae	Mukerjee (1978)
Yersinia enterocolitica	Bergan (1978c)

Phage types may be modified by genetic mechanisms including lysogenic conversion, loss of prophages, and acquisition or loss of R plasmids. In addition, there is a continuous requirement to maintain the typing set of bacteriophages in a viable state by regular serial passage.

It can be concluded that phage typing is a highly sensitive, but technically demanding, method of typing which can provide valuable information for certain species. Larger laboratories will be able to undertake phage typing of particular species on a routine basis, but smaller laboratories will probably find it easier to send isolates to a reference laboratory. It must, however, be emphasized that sets of phages for typing purposes have been developed and refined for particular species over several decades. This is not a technique which can be applied readily to a 'new' organism or application in response to a sudden emerging problem; indeed, for some genera no phages have yet been isolated.

1.2.3 SEROTYPING

Serotyping is one of the oldest typing procedures, but particularly when used in conjunction with other typing methods, still represents an important tool for routine identification and typing of many microbial species (Table 1.3). Based on reactions with specific antisera (e.g. antisomatic, antiflagellar or anticapsular), raised according to the antigenic structure of microbes in each group, the method has been developed in detail for bacteria belonging to the Enterobacteriaceae (Ewing, 1986), and is also a method which has received considerable attention for typing *P. aeruginosa* (Lányi and Bergan, 1978). Serotyping has the advantage that it can be applied to many different genera, although in many cases a given set of reagents can be applied only to a single species. As with phage typing, it may require some time to develop a serotyping scheme for a 'new' application, since the development of such a scheme is quite a complex process, with requirements for immunization of rabbits with specific purified antigens, agglutination tests, immunization of further rabbits with unreactive strains, and cross-absorption experiments. Although some possible changes in structural antigens, related particularly to lysogenic conversion, have been noted (Meitert and Meitert, 1978), in general it seems that the serotype of a microbe is a relatively stable and reliable typing marker. The main disadvantages seem to be associated with problems in antisera production and standardization of methodology. It is also appropriate to note that, since the method depends on the production of a range of specific antisera, a task which is logistically difficult for a routine microbiology laboratory to perform, serotyping schemes for certain genera may normally be available only in reference laboratories.

1.2.4 BACTERIOCIN TYPING

Bacteriocins are bactericidal substances, normally proteins, which are active against different strains of bacteria. Bacteriocin typing is performed by testing the sensitivity of 'unknown' strains to bacteriocins produced by a set of standard selected strains. The method can also be performed in reverse by testing the ability of 'unknown' strains to produce bacteriocins with activity against a standard set of indicator strains. In general, a strain producing a particular bacteriocin is also resistant to its action. As with serotyping, bacteriocin typing has been applied particularly to *P. aeruginosa* and members of the Enterobacteriaceae (Table 1.4). Although production of bacteriocins or sensitivity to them are relatively stable properties, it is known that these properties may be encoded by transmissible R plasmids. Additional disadvantages are that the method is relatively labour-intensive, particularly when cross-streaking techniques are used, and may require considerable development work before it can be used for a 'new' application.

Table 1.3 Examples of microorganisms for which serotyping schemes have been developed

Organism	Reference
Acinetobacter spp.	Traub (1989, 1990)
Actinobacillus spp.	McArthur *et al.* (1986)
Aeromonas spp.	Guinée and Jansen (1987)
Bacillus spp.	Berkeley *et al.* (1984)
Bordetella pertussis	Robinson *et al.* (1989)
Brucella spp.	McCullough (1978)
Campylobacter spp.	Mills *et al.* (1991)
Candida albicans	Brawner (1991)
Chlamydia psittaci	Anderson (1991)
Chlamydia trachomatis	Wang and Grayston (1991)
Citrobacter spp.	Lányi (1984)
Clostridium difficile	Toma *et al.* (1988)
Clostridium perfringens	Gross *et al.* (1989)
Cucumber mosaic virus	Binahmad and Scott (1985)
Dengue virus	Nawa *et al.* (1985)
Edwardsiella tarda	Tamura *et al.* (1988)
Enterobacter cloacae	Verschraegen *et al.* (1988)
Erwinia spp.	Slade and Tiffin (1984)
Erysipelothrix rhusiopathiae	Norrung *et al.* (1987)
Escherichia coli	Ørskov and Ørskov (1984)
Gardnerella vaginalis	Ison *et al.* (1987)
Haemophilus influenzae	Himmelreich *et al.* (1985)
Herpes simplex virus	Gleaves *et al.* (1985)
Human immunodeficiency virus	Baillou *et al.* (1991)
Klebsiella spp.	Ørskov and Ørskov (1984)
Leptospira spp.	Dikken and Kmety (1978)
Listeria monocytogenes	McLauchlin *et al.* (1986)
Maize streak virus	Pinner and Markham (1990)
Microsporum canis	Polonelli *et al.* (1986)
Moraxella bovis	Moore and Lepper (1991)
Neisseria gonorrhoeae	Moyes and Young (1991)
Neisseria meningitidis	Wedege *et al.* (1990)
Pasteurella haemolytica	Sanchis *et al.* (1988)
Pasteurella multocida	Ireland *et al.* (1989)
Plasmodium falciparum	McBride *et al.* (1984)
Plesiomonas shigelloides	Aldova (1987)
Proteus spp.	Larsson (1984)
Pseudomonas aeruginosa	Burke *et al.* (1990)
Rotaviruses	Coulson *et al.* (1987)
Salmonella spp.	Lindberg and Le Minor (1984)
Serratia liquefaciens	Traub (1991)
Serratia marcescens	Gaston and Pitt (1989)
Staphylococcus aureus	Hochkeppel *et al.* (1987)
Streptococcus faecalis	Smyth *et al.* (1987)
Streptococcus mutans	Ota *et al.* (1984)
Streptococcus pneumoniae	Waltman *et al.* (1988)
Group B streptococci	Pasaribu *et al.* ((1985)
Group G streptococci	Lawal and Dosunmuogunbi (1986)

Table 1.3 *continued*

Treponema hyodysenteriae	Li *et al.* (1991)
Vibrio anguillarum	Sorensen and Larsen (1986)
Vibrio cholerae	Gustafsson (1984)
Vibrio fluvialis/furnissii	Shimada *et al.* (1991)
Yersinia enterocolitica	Aleksic and Bockemuhl (1984)
Yersinia ruckeri	Davies (1990)

1.2.5 CONCLUSIONS

From the above descriptions it can be seen that none of the conventional biological typing methods offers an ideal approach for the subdividing of microbial species. The importance and applicability of each of the methods may vary from one species to another, and also according to the precise geographical location. For many species, combined use of the different methods may offer the best approach, but such an approach may not be possible, or may take a long time to develop, for organisms which have not been well studied previously. Progress in molecular biology has resulted in the availability of methods which have the potential to be used to study diversity in any microbial species. The remainder of this volume is devoted to the consideration of these molecular methods, together with the possibility that they may provide a 'universal' approach to microbial identification and typing.

Table 1.4 Examples of microorganisms for which bacteriocin typing schemes have proved useful

Organism	Reference
Acinetobacter	Andrews (1986)
Actinomyces pyogenes	Lammler (1990)
Clostridium difficile	Bacon *et al.* (1988)
Clostridium perfringens	Watson (1985)
Enterobacteriaceae	Gillies (1978)
Klebsiella pneumoniae	Chhibber *et al.* (1988)
Morganella morgani	Senior (1987)
Pseudomonas aeruginosa	Govan (1978)
Pseudomonas cepacia	Govan and Harris (1985)
Serratia liquefaciens	Traub (1991)
Serratia marcescens	Traub (1978)
Staphylococci	Skalka (1987)
Streptococci: group A	Tagg and Martin (1984)
β-haemolytic	Ragland and Tagg (1990)
Vibrio cholerae	Brandis (1978)

1.3 MOLECULAR APPROACHES TO MICROBIAL IDENTIFICATION AND TYPING

As indicated in the previous sections, conventional biological typing procedures have been devised only after long-term efforts, and are available for only a limited range of microbial species. So far as a 'universal' molecular approach to microbial identification and typing is concerned, nucleic acids, proteins and lipopolysaccharides are the only macromolecules which carry enough information in their sequences to allow a simple uniform approach to the study of microbial diversity and the development of typing methods. Historically, methods used to isolate and characterize macromolecules have involved complex and time-consuming procedures which have prevented their introduction into routine microbiology laboratories. More recently, such techniques have become increasingly rapid and simple, to the extent that they are now readily accessible to most trained microbiologists. The main concepts of the available methods are outlined below and are discussed in detail in the succeeding chapters.

1.3.1 NUCLEIC ACID ANALYSIS

Although the examination of proteins and lipopolysaccharides has been used successfully to type a variety of different microbes (sections 1.3.2 and 1.3.3), all such techniques suffer from the limitation that they analyse the phenotype rather than the genotype of a particular organism. From a purist's point of view, it is preferable to analyse the genotype of an organism since this type of analysis does not rely on the expression of particular genes encoding proteins or lipopolysaccharides, and is therefore not subject to phenotypic variation. Methods for the analysis of nucleic acids can be divided broadly into those which examine the size and structure of individual nucleic acid molecules, and those which examine the degree of relatedness between molecules by hybridization procedures.

(a) Analysis of plasmid DNA

The easiest nucleic acid molecules to examine are plasmids. The publication by Birnboim and Doly (1979) of a simple rapid method for isolating plasmid DNA, followed by the formulation of numerous derivative methods with minor modifications for particular genera, now means that plasmid isolation and analysis can be performed in any laboratory that possesses a high-speed bench centrifuge, simple agarose gel electrophoresis equipment and an ultraviolet trans-illuminator. Simple plasmid pattern analysis has already been applied to the typing of many different bacterial genera (Chapter 2); however, the possible long-term lack of stability of plasmids in particular strains may pose a major drawback, and plasmid analysis can obviously not be used if the strains being investigated do not contain plasmids to begin with.

More detailed analysis of plasmid DNA, particularly large plasmid molecules, can be performed following restriction endonuclease digestion of purified plasmid DNA. Restriction endonucleases are enzymes that recognize specific base pair sequences on DNA molecules, and then cleave these sequences at a defined position to produce a 'fingerprint' of linear DNA fragments following electrophoresis on an agarose gel. This simple procedure generates much additional information and allows unrelated plasmids of similar size to be distinguished readily.

Individual DNA fragments (or whole plasmids) can also be compared by means of DNA–DNA hybridization procedures, normally following transfer of DNA from the gel to a membrane support before hybridization.

(b) Analysis of chromosomal DNA

The problem of using nucleic acids to type isolates that either lack plasmids or carry unstable plasmids can be overcome by examining chromosomal DNA restriction fragments.

The microbial chromosome is a large and fragile molecule which requires special isolation techniques to avoid random cleavage by mechanical forces during its isolation. Procedures have now been developed in which intact cells embedded in agarose are lysed and deproteinized *in situ*. The agarose protects the DNA from shear forces, and enables the DNA sample to be manipulated readily. Following treatment with appropriate restriction endonucleases, samples in plugs of agarose can be loaded directly into the wells of agarose gels for subsequent electrophoresis (Schwartz and Cantor, 1984).

An additional complication arises from the fact that conventional electrophoresis techniques do not permit resolution of DNA fragments larger than about 50 kb. Restriction enzymes which cut whole microbial chromosomes into fragments smaller than 50 kb tend to generate large numbers of similar-sized fragments which are difficult to separate and interpret. This problem has been overcome by the development of **pulsed-field gel electrophoresis** (PFGE) techniques in which, by alternating the electric field between spatially distinct pairs of electrodes, DNA molecules of up to 10 megabases are able to re-orientate and move differentially through the pores in an agarose gel (Carle and Olson, 1984; Schwartz and Cantor, 1984; Chu, Vollrath and Davis, 1986). This allows the use of restriction enzymes which generate a relatively small number of fragments from a whole chromosome.

Several different pulsed-field systems are available (Chapter 2), but perhaps the most popular is based on the CHEF (clamped homogeneous electric fields) technique in which a uniform electric field is generated at all points on a gel, with an optimal 120° angle of alternating pulses, by means of an hexagonal array of 24 electrodes (Carle and Olson, 1984). Directly visualized restriction enzyme digests of chromosomal DNA produced by systems of this type can be used to generate reproducible DNA 'fingerprints' for a wide range of microorganisms. These fingerprints are highly discriminatory for different strains, and allow similarities and differences between strains to be determined readily.

(c) Hybridization procedures

It is now commonplace to identify the presence/absence of particular microbial species in a clinical or environmental sample by means of hybridization with specific nucleic acid probes. While an increasing number of kits are now becoming available commercially for such tests, in their basic form such tests are an identification rather than a typing method. They also suffer from the drawback that each individual species to be identified requires its own specific probe. However, it has been possible to develop typing procedures by combining DNA hybridization with restriction endonuclease analysis. Restriction fragments of total chromosomal DNA separated on an agarose gel can be transferred to nitrocellulose or nylon membranes and then hybridized with labelled probes consisting of randomly cloned DNA from the same genomic species (Tompkins *et al.*, 1986). Each of the cloned fragments of chromosomal DNA can be used in turn to identify identical sequences in the restriction digest pattern of the organism being typed. Identical or closely related strains will have a large number of hybridizing fragments in common, while unrelated strains will have none or very few hybridizing fragments in common.

A generalized typing method ('**ribotyping**') based on the above procedure has been suggested by Grimont and Grimont (1986). This 'universal' approach relies on the use of ribosomal RNA (rRNA), or the corresponding genes, as a probe. rRNA sequences contain conserved regions which can hybridize to the rRNA genes of any microbial species. Depending on the precise probe used, some microbial species are so genetically homogeneous that all strains examined give a single rRNA gene restriction pattern following hybridization, while, with the same probe, many different combinations of patterns may be observed with strains belonging to other species.

The various DNA–DNA and DNA–rRNA hybridization techniques and formats that are available are discussed in detail in Chapter 3, but all are similar in so far as they are fast, conceptually simple, relatively inexpensive and universally applicable.

(d) Amplification procedures and sequencing techniques

The basis of all sequencing techniques is the generation of four nested series of labelled DNA fragments, each with one common end and the other end generated in a specific manner according to one of the four bases which form a nucleic acid molecule. The four sets of fragments are separated in parallel by electrophoresis on a high resolution acrylamide slab gel. The sequence can then be determined from the relative order of the labelled fragments in each of the four tracks on the gel (Chapter 4).

Two techniques for generating the specific fragments have gained wide acceptance: the 'chemical' (Maxam and Gilbert, 1977) and 'chain termination' (Sanger *et al.*, 1977) methods. These techniques are described fully in Chapter 4, but at this point it is sufficient to say that, for microbial identification purposes, sequence determination is usually performed by the chain termination method since it is less time-consuming and simpler than the chemical method.

Three methods are used commonly to isolate nucleic acids and obtain sufficient material for sequencing. First, specific genes can be cloned into multicopy plasmid vectors by means of recombinant DNA techniques. Secondly, RNA transcripts of particular genes can be isolated and sequenced. Thirdly, target DNA sequences can be amplified and sequenced directly from whole genomic DNA. A variety of techniques (Chapter 4), including the polymerase chain reaction (PCR; Mullis and Faloona, 1987), the Qβ method (Kramer *et al.*, 1991) and other amplification methods (Van Brunt, 1990), all enable up to a million-fold amplification of the DNA or RNA sequence of interest from a single copy present in a whole genomic sample of nucleic acid. This can then be used as a source of material for sequencing reactions and typing purposes, or can often be sufficiently diagnostic that, for identification purposes, it may be sufficient simply to recognize it by its size on agarose gels. However, better specificity and sensitivity can be obtained by examining the products of an amplification reaction by hybridization with a specific probe.

Identification of a specific sequence of nucleic acid is a valuable diagnostic tool because of the high specificity of this approach. Although nucleic acid sequencing is a comparatively new approach to microbial identification, and is certainly not at present a typing technique for routine use, advances in sequencing technology during recent years have been so rapid that comparative sequencing of homologous genes is now a standard technique in molecular biology. The technique is especially valuable when conserved genes (e.g. those encoding rRNA) are sequenced to establish evolutionary and taxonomic relationships between different bacteria.

1.3.2 PROTEIN ANALYSIS

Proteins perform a range of structural and metabolic functions in prokaryotes, eukaryotes and viruses. The diversity and number of proteins produced – up to 2000 for some bacteria – has provided ample scope for the development of typing and identification schemes based on comparative analysis of protein or polypeptide profiles. This approach has been applied successfully to bacteria for identification of organisms to the species level and, particularly, for subtyping within bacterial species. In general, profiles of major proteins produced by strains of the same species are very similar, but differ significantly from those of other species. Within a species, minor differences between strains form the basis for subtyping schemes. Although not used as extensively as for bacteria, protein analysis has been used also for the differentiation and subtyping of fungi and some viruses.

As indicated above (section 1.3.1), protein expression is a phenotypic character that is regulated by both the genotype of the organisms being studied and environmental factors. In all comparative analyses based on protein profiles, growth conditions for preparation of test organisms should, therefore, be standardized as far as is possible in order to avoid artefacts and increase reproducibility.

(a) Protein profiles

Generation of protein profiles involves solubilization of microbial proteins, either from whole cells or subcellular fractions, and separation of the solubilized material by electrophoresis in gel matrices such as polyacrylamide or starch. Generally, these techniques use relatively small numbers of microbial cells, and many strains can be compared simultaneously.

Depending on the type of electrophoresis system used, proteins are separated on the basis of a number of physicochemical properties, including size, charge and isoelectric point. Separated proteins, or their component subunits, are then detected by staining techniques to generate a characteristic protein profile. Profiles may be analysed by eye, but more complex patterns may require computer-assisted analysis.

In one of the techniques used most widely, SDS-PAGE (sodium dodecylsulphate polyacrylamide gel electrophoresis), proteins are denatured by heat, detergent and reducing agents to produce polypeptide subunits which are separated by electrophoresis on the basis of molecular mass. Alternatively, proteins can be solubilized and electrophoresed under conditions that retain native protein structure and biological activity. The very powerful technique of multilocus enzyme electrophoresis (MEE) involves separation of proteins with enzymic activity, which are then detected on gels with specific substrates and chromogens.

(b) Antigenic analysis of proteins

Most microbial proteins are antigenic, and serotyping methods, based on protein antigens, which use agglutination or precipitation techniques have been employed widely in the past. Although these traditional methods are still used extensively in medical microbiology (section 1.2.3), the development of electrophoresis techniques that allow detailed antigenic analysis of individual proteins, or groups of proteins, has provided even more sensitive means of typing for some groups of microorganisms. Minor structural differences between proteins, which may not be detectable by conventional staining techniques, can be identified by means of specific antibodies. These may be polyclonal, recognizing many different epitopes on a particular protein, or monoclonal, recognizing a single antigenic epitope. By testing a panel of antibodies for reactivity with a particular group of organisms, typing schemes can be devised, based on the presence or absence of single antigenic epitopes, that employ these methodologies.

1.3.3 LIPOPOLYSACCHARIDE ANALYSIS

Lipopolysaccharide (LPS), or endotoxin, is a component of the outer-membrane of Gram-negative bacteria. LPS is a complex molecule which is anchored into the bacterial outer-membrane by a hydrophobic moiety called lipid A. Linked to lipid A is a short oligosaccharide chain that forms the core of the molecule. Attached to the core in some bacteria are long polysaccharide side-chains of variable composition and antigenicity. LPS of this type is called 'smooth', in contrast to 'rough' LPS which lacks the polysaccharide side-chains. Another rough variant of LPS, termed lipo-oligosaccharide (LDS), is present in some bacteria such as *Neisseria gonorrhoeae*. Typing schemes based on LPS analysis are dependent on differences in electrophoretic mobility, or antigenicity, associated with variations in the composition of the LPS molecule, and particularly with the presence, absence or length of the polysaccharide side-chains.

(a) LPS profiles

In order to analyse LPS, this molecule must first be separated from other bacterial components, particularly proteins, which may interfere with or mask LPS profiles on polyacrylamide gels. LPS can be extracted from bacteria by means of phenol/chloroform treatment, but for small-scale analytical work, LPS prepared by protease digestion of crude bacterial lysates is often used. The protease-resistant LPS molecules are then separated by electrophoresis on polyacrylamide gels, primarily on the basis of size and charge, and detected with silver-staining techniques that recognize sugar residues present in the core and polysaccharide side-chains. Characteristic banding patterns are found for rough and smooth LPS types, and minor differences in LPS composition can be detected by means of serological techniques which employ monoclonal or polyclonal antibodies.

1.3.4 OTHER MOLECULAR METHODS INVOLVING CHEMICAL ANALYSIS

(a) Analysis by gas–liquid chromatography (GLC)

Gas–liquid chromatography is based on the passage through a packed column (the stationary phase) of a carrier gas, usually oxygen-free nitrogen, argon or helium, at a pre-determined rate. The choice of stationary phase is critical and depends on the separation to be attempted. Examples of the many hundreds of available stationary phases are listed by Drucker (1976). The column is maintained at a specified temperature and a suitable injection system is used to apply a sample, either as a solid, liquid or gas, onto the column. Substances eluted from the far end of the column in the gas phase pass through a sensitive detector system, normally linked to a chart recorder. The versatility of this technique, in terms of the range of compounds that can be analysed, has resulted in its use in two major areas for microbial identification.

First, GLC has been used for the analysis of fermentation end-products, particularly those produced by anaerobic organisms which tend to use organic compounds as terminal electron acceptors, thereby reducing them to acids, alcohols and ketones, all of which may have taxonomic potential. Indeed, the definitive identification of anaerobes relies on the use of biochemical reactions in combination with GLC analysis of metabolic end-products (Dowell and Hawkins, 1977; Holdeman, Cato and Moore, 1977; 1987; Sutter *et al.*, 1985). As an example, while production of propionic acid is a key identification character for members of the propionibacteria, its routine identification is made much simpler with the aid of GLC.

Secondly, GLC can be used to obtain a profile of the cellular components of microbes. This technique has generally involved analysis of volatile derivatives of cellular components such as amino acids, fatty acids or sugars, and can be applied equally well to aerobes and facultative anaerobes as well as obligate anaerobes. Analysis of metabolic short-chain fatty acids has received particular attention, and a commercial computer-assisted microbial identification system is available in which GLC fatty acid profiles are compared with a reference data base. The rapid identification results obtained with this system have been shown to correlate reasonably well with results obtained by more laborious conventional techniques for Gram-negative anaerobic bacteria (Stoakes *et al.*, 1991). Other examples of the use of GLC for microbial identification purposes are listed in Table 1.5.

Table 1.5 Examples of the use of GLC for the identification of specific microorganisms

Organism	Reference
Campylobacter spp.	Heinzer (1985)
Candida spp.	Maliwan *et al.* (1984)
Clostridium butyricum	Magot *et al.* (1983)
Clostridium difficile	Johnson *et al.* (1989)
Haemophilus influenzae	Bohning *et al.* (1989)
Legionella spp.	Marmet *et al.* (1988)
Listeria spp.	Ninet *et al.* (1992)
Mycobacterium tuberculosis	Maliwan *et al.* (1988)
Pseudomonas spp.	Stead (1992)
Saccharomyces spp.	Augustyn *et al.* (1991)
Staphylococci	Magee *et al.* (1983)

The main drawbacks of GLC are the high capital costs of the apparatus and the large number of technical parameters that have to be considered when designing an analytical strategy (Drucker, 1976). Nevertheless, if standardized semi-automated procedures can be employed, the technique can be an invaluable aid to the identification of taxonomically complex groups of microorganisms. The technique is not yet suitable for use as a generalized typing procedure.

(b) Analysis by pyrolysis mass spectrometry

This method is based on the principle that pyrolysis of pure cultures generates mass ion spectra which are characteristic of the individual organism. A small sample of culture is placed on a sample foil and loaded into a pyrolysis mass spectrometer. The sample is pyrolysed under vacuum and the composition of the evolved gas (the pyrolysate) is recorded to generate a 'fingerprint' of the starting material. The system can be entirely automated, with sample loading, extraction and analysis all under computer control. Pyrolysis mass spectrometry has been used to provide inter-strain comparisons for a variety of microorganisms (Table 1.6), with computerized analysis of the data being used to select those mass ions which provide the best discrimination between the various spectra. These then form the basis of a mathematically derived analysis of similarities and differences which can be used for interstrain comparison.

Table 1.6 Examples of microorganisms in which pyrolysis mass spectrometry has been used to provide inter-strain comparisons

Organism	Reference
Candida albicans	Magee *et al.* (1988)
Corynebacterium spp.	Hindmarch *et al.* (1990)
Escherichia coli	Goodacre and Berkeley (1990)
Fusobacterium spp.	Magee *et al.* (1989)
Legionella pneumophila	Sisson *et al.* (1991b)
Listeria monocytogenes	Freeman *et al.* (1991c)
Mycobacterium tuberculosis	Sisson *et al.* (1991c)
Pseudomonas aeruginosa	Sisson *et al.* (1991a)
Rhizobium meliloti	Goodacre *et al.* (1991)
Staphylococcus aureus	Gould *et al.* (1991)
Staphylococcus epidermidis	Freeman *et al.* (1991b)
Streptococcus pneumoniae	Freeman *et al.* (1991a)
Xanthomonas maltophilia	Orr *et al.* (1991)

It is important to note that pyrolysis mass spectrometry is not really a universal typing method since spectra can vary according to the duration and conditions of incubation of the organisms being studied. However, when a set of organisms is prepared under identical conditions and analysed in a single batch, comparisons have been shown to be objective and highly discriminatory, allowing the identification of closely-related groups ('types') of isolates within a larger collection (Gutteridge and Norris, 1979; Freeman *et al.*, 1990). Although the consumable costs are less than £1 per sample, the high capital equipment costs of c. £80 000 seem destined to ensure that the method remains as an interesting and effective research tool, or is confined to a few major reference laboratories, rather than becoming a widespread and accepted typing technique.

REFERENCES

Aldova, E, (1987) Serotyping of *Plesiomonas shigelloides* strains with our own antigenic scheme – an attempted epidemiological study. *Zentralblatt für Bakteriologie Mikrobiologie und Hygiene Series A*, **265**, 253–62.

Aleksic, S. and Bockemuhl, J. (1984) Proposed revision of the Wauters *et al* antigenic scheme for serotyping of *Yersinia enterocolitica*. *Journal of Clinical Microbiology*, **20**, 99–102.

Anderson, A.A. (1991) Serotyping of *Chlamydia psittaci* isolates using serovar-specific monoclonal antibodies with the microimmunofluorescence test. *Journal of Clinical Microbiology*, **29**, 707–11.

Andrews, H.J. (1986) *Acinetobacter* bacteriocin typing. *Journal of Hospital Infection*, **7**, 169–75.

Augustyn, O.P.H., Ferreira, D. and Kock, J.L.F. (1991) Differentiation between yeast species, and strains within a species, by cellular fatty-acid analysis. 4. *Saccharomyces sensu-stricto, Hanseniaspora, Saccharomycodes* and *Wickerhamiella*. *Systematic and Applied Microbiology*, **14**, 324–34.

Bacon, A.E., Fekety, R., Schaberg, D.R. and Faix, R.G. (1988) Epidemiology of *Clostridium difficile* colonization in newborns – results using a bacteriophage and bacteriocin typing system. *Journal of Infectious Diseases*, **158** 349–54.

Baillou, A., Janvier, B., Leonard, G. *et al.* (1991) Fine serotyping of human-immunodeficiency virus serotype-1 (HIV-1) and HIV-2 infections by using synthetic oligopeptides representing an immunodominant domain of HIV-1 and HIV-2 simian immunodeficiency virus. *Journal of Clincal Microbiology*, **29**, 1387–91.

Bergan, T. (1978a) Phage-typing of *Pseudomonas aeruginosa*, in *Methods in Microbiology*, vol. 10 (eds) T. Bergan & J.R. Norris. Academic Press, London, pp. 169–99.

Bergan, T. (1978b) Phage-typing of *Proteus*, in *Methods in Microbiology*, vol. 11 (eds T. Bergan and J.R. Norris), Academic Press, London, pp. 243–58.

Bergan, T. (1978c) Bacteriophage typing of *Yersinia enterocolitica*, in *Methods in Microbiology*, vol. 12 (eds T. Bergan and J.R. Norris), Academic Press, London, pp. 25–36.

Bergan, T. (1979) Bacteriophage typing of *Shigella*, in *Methods in Microbiology*, vol. 13 (eds T. Bergan and J.R. Norris), Academic Press, London pp. 178–286.

Berkeley, R.C.W., Logan, N.A., Shute, L.A. and Capey, A.C. (1984) Identification of *Bacillus* species, in *Methods in Microbiology*, vol. 16 (ed T. Bergan), Academic Press, London, pp. 292–328.

Biberslein, E.L. (1978) Biotyping and serotyping of *Pasteurella haemolytica*, in *Methods in Microbiology*, vol. 10 (eds T. Bergan and J.R. Norris), Academic Press, London, pp. 253–69.

Binahmad, I. and Scott, H.A. (1985) Identification and serotyping of cucumber mosaic and peanut stunt viruses from Arkansas. *Plant Disease*, **69**, 891–3.

Birnboim, H.D. and Doly, J. (1979) A rapid alkaline extraction procedure for screening recombinant plasmid DNA. *Nucleic Acids Research*, **7**, 1513–23.

Bohning, A., Steinbach, K. and Mutters, R. (1989) Rapid identification of *Haemophilus influenzae* serovar b by gas–liquid chromatography using carbohydrate fingerprints. *Zentralblatt für Bakteriologie International Journal of Medical Microbiology*, **272**, 19–29.

Bouvet, P.J.M. (1991) Typing of *Acinetobacter*, in *The Biology of Acinetobacter: Taxonomy, Clinical Importance, Molecular Biology, Physiology, Industrial Relevance* (eds K.J. Towner, E. Bergogne-Bérézin and C.A. Fewson), Plenum Press, New York, pp. 37–51.

Brandis, H. (1978) Vibriocin typing, in *Methods in Microbiology*, vol. 12 (eds T. Bergan and J.R. Norris), Academic Press, London, pp. 117–26.

Brawner, D.L. (1991) Comparison between methods for serotyping of *Candida albicans* produces discrepancies in results. *Journal of Clinical Microbiology*, **29**, 1020–5.

Burke, V., Richardson, C.J.L. and Robinson, J. (1990) Serotype and serum sensitivity of *Pseudomonas aeruginosa* from children with cystic fibrosis – longitudinal studies and typing with monoclonal antibodies. *Pathology*, **22**, 223–6.

Carle, G.F. and Olson, M.V. (1984) Separation of chromosomal DNA molecules from yeast by orthogonal field alternation gel electrophoresis. *Nucleic Acids Research*, **12**, 5647–64.

Catlin, B.W. (1978) Characteristics and auxotyping of *Neisseria gonorrhoaeae*, in *Methods in Microbiology*, vol. 10 (eds T. Bergan and J.R. Norris), Academic Press, London, pp. 345–80.

Chhibber, S., Goel, A., Kapoor, N. *et al.* (1988) Bacteriocin (Klebocin) typing of clinical isolates of *Klebsiella pneumoniae*. *European Journal of Epidemiology*, **4**, 115–8.

Chu, G., Vollrath, D. and Davis, R. (1986) Separation of large DNA molecules by contour-clamped homogeneous electric fields. *Science*, **234**, 1582–5.

Corbel, M.J. (1984) Phage typing of *Brucella*, in *Methods in Microbiology*, vol. 16 (ed T. Bergan), Academic Press, London, pp. 23–74.

Coulson, B.S., Unicomb, L.E., Pitson, G.A. and Bishop, R.F. (1987) Simple and specific enzyme-immunoassay using monoclonal-antibodies for serotyping human rotaviruses. *Journal of Clinical Microbiology*, **25**, 509–15.

Davies, R.L. (1990) O-Serotyping of *Yersinia ruckeri* with special emphasis on European isolates. *Veterinary Microbiology*, **22**, 299–307.

Dikken, H. and Kmety, E. (1978) Serological typing methods of leptospires, in *Methods in Microbiology*, vol. 11 (eds T. Bergan and J.R. Norris), Academic Press, London, pp. 260–307.

Dowell, V.R. and Hawkins, T.M. (1977) *Laboratory Methods in Anaerobic Bacteriology, CDC Laboratory Manual*. Center for Disease Control, Atlanta.

Drucker, D.B. (1976) Gas–liquid chromatographic chemotaxonomy, in *Methods in Microbiology*, vol. 9, (ed J.R. Norris), Academic Press, London, pp. 52–125.

Ewing, W.H. (1986) *Edwards and Ewing's Identification of Enterobacteriaceae*, 4th edn, Elsevier Science, New York.

Fanti, F., Conti, S., Campani *et al.* (1989) Studies on the epidemiology of *Aspergillus fumigatus*. *European Journal of Epidemiology*, **5**, 8–14.

Freeman, R., Goodfellow, M., Gould, F.K. *et al.* (1990) Pyrolysis mass spectrometry (Py-Ms) for the rapid epidemiological typing of clinically significant bacterial pathogens. *Journal of Medical Microbiology*, **32**, 283–6.

Freeman, R., Gould, F.K., Sisson, P.R. and Lightfoot, N.F. (1991a) Strain differentiation of capsule type 23 penicillin-resistant *Streptococcus pneumoniae* from nosocomial infections by pyrolysis mass spectrometry. *Letters in Applied Microbiology*, **13**, 28–31.

Freeman, R., Gould, F.K., Wilkinson, R. *et al.* (1991b) Rapid inter-strain comparison by pyrolysis mass spectrometry of coagulase-negative staphylococci from persistent CAPD peritonitis. *Epidemiology and Infection*, **106**, 239–46.

Freeman, R., Sisson, P.R., Lightfoot, N.F. and McLauchlin, J. (1991c) Analysis of epidemic and sporadic strains of *Listeria monocytogenes* by pyrolysis mass spectrometry. *Letters in Applied Microbiology*, **12**, 133–6.

Gaston, M.A. and Pitt, T.L. (1989) Improved O-serotyping method for *Serratia marcescens*. *Journal of Clinical Microbiology*, **27**, 2702–5.

Gillies, R.R. (1978) Bacteriocin typing of Enterobacteriaceae, in *Methods in Microbiology*, vol. 11 (eds T. Bergan and J.R. Norris), Academic Press, London, pp. 79–86.

Gleaves, C.A., Wilson, D.J., Wold, A.D. and Smith, T.F. (1985) Detection and serotyping of herpes-simplex virus in MRC–5 cells by use of centrifugation and monoclonal antibodies 16 h postinoculation. *Journal of Clinical Microbiology*, **21**, 29–32.

Goodacre, R. and Berkeley, R.C.W. (1990) Detection of small genotypic changes in *Escherichia coli* by pyrolysis mass spectrometry. *FEMS Microbiology Letters*, **71**, 133–8.

Goodacre, R., Hartmann, A., Beringer, J.E. and Berkeley, R.C.W. (1991) The use of pyrolysis mass spectrometry in the characterization of *Rhizobium meliloti*. *Letters in Applied Microbiology*, **13**, 157–60.

Goodfellow, M. and O'Donnell, A.G. (1989) Search and discovery of industrially-significant actinomycetes, in *Microbial Products: New Approaches* (eds S. Baumberg, I.S. Hunter and P.M. Rhodes), Cambridge University Press, Cambridge, pp. 343–83.

Gould, F.K., Freeman, R., Sisson, P.R., Cookson, B.D. and Lightfoot, N.F. (1991) Inter-strain comparison by pyrolysis mass spectrometry in the investigation of *Staphylococcus aureus* nosocomial infection. *Journal of Hospital Infection*, **19**, 41–8.

Govan, J.R.W. (1978) Pyocin typing of *Pseudomonas aeruginosa*, in *Methods in Microbiology*, vol. 10 (eds T. Bergan and J.R. Norris), Academic Press, London, pp. 61–91.

Govan, J.R.W. and Harris, G. (1985) Typing of *Pseudomonas cepacia* by bacteriocin susceptibility and production. *Journal of Clinical Microbiology*, **22**, 490–4.

Grimont, F. and Grimont, P.A.D. (1986) Ribosomal ribonucleic acid gene restriction patterns as potential taxonomic tools. *Annales de l'Institut Pasteur Microbiologie*, **137B**, 165–75.

Gross, T.P., Kamara, L.B., Hatheway, C.L. *et al.* (1989) *Clostridium perfringens* food poisoning – use of serotyping in an outbreak setting. *Journal of Clinical Microbiology*, **27**, 660–3.

Guinée, P.A.M. and Jansen, W.H. (1987) Serotyping of *Aeromonas* species using passive hemagglutination. *Zentralblatt fur Bakteriologie Mikrobiologie und Hygiene Series A*, **265**, 305–13.

Guinée, P.A.M. and van Leeuwen, W.J. (1978) Phage-typing of *Salmonella*, in *Methods in Microbiology*, vol. 11 (eds T. Bergan and J.R. Norris), Academic Press, London, pp. 157–91.

Gustafsson, B. (1984) Monoclonal antibody-based enzyme-linked immunosorbent assays for identification and serotyping of *Vibrio cholerae* 01. *Journal of Clinical Microbiology*, **20**, 1180–5.

Gutteridge, C.S. and Norris, J.R. (1979) The application of pyrolysis techniques to the identification of microorganisms. *Journal of Applied Bacteriology*, **47**, 5–43.

Hawkey, P.M. (1989) Laboratory investigation of nosocomial infection, in *Medical Bacteriology: A Practical Approach* (eds P.M. Hawkey and D.A. Lewis), IRL Press, Oxford, pp. 247–69.

Heinzer, J. (1985) Gas–liquid chromatography for differentiation of thermophilic campylobacters and *Campyolobacter pyloridis*. *Experientia*, **41**, 548.

Himmelreich, C.A., Barenkamp, S.J. and Storch, G.A. (1985) Comparison of methods for serotyping isolates of *Haemophilus influenzae*. *Journal of Clinical Microbiology*, **21**, 158–60.

Hindmarch, J.M., Magee, J.T., Hadfield, M.A. and Duerden, B.I. (1990) A pyrolysis mass spectrometry study of *Corynebacterium* spp. *Journal of Medical Microbiology*, **31**, 137–49.

Hochkeppel, H.K., Braun, D.G., Vischer, W. *et al.* (1987) Serotyping and electron-microscopy studies of *Staphylococcus aureus* clinical isolates with monoclonal antibodies to capsular polysaccharide type–5 and type–8. *Journal of Clinical Microbiology*, **25**, 526–30.

Holdeman, L.V., Cato, E.P. and Moore, W.E.C. (1977) *Anaerobe Laboratory Manual*, Virginia Polytechnic Institute and State University, Blacksburg.

Holdeman, L.V., Cato, E.P. and Moore, W.E.C. (1987) *Anaerobe Laboratory Manual Update*, Virginia Polytechnic Institute and State University, Blacksburg.

Ireland, L.A., Milner, A.R. and Smart, I.J. (1989) Serotyping of isolates of *Pasteurella multocida* from chickens. *Australian Veterinary Journal*, **66**, 119–20.

Ison, C.A., Harvey, D.G., Tanna, A. and Easmon, C.S.F. (1987) Development and evaluation of scheme for serotyping *Gardnerella vaginalis*. *Genitourinary Medicine*, **63**, 196–201.

Jain, R.K., Burlage, R.S. and Sayler, G.S. (1988) Methods for detecting recombinant DNA in the environment. *CRC Critical Reviews of Biotechnology*, **8**, 33–84.

Johnson, L.L., McFarland, L.V., Dearing, P. *et al.* (1989) Identification of *Clostridium difficile* in stool specimens by culture-enhanced gas–liquid chromatography. *Journal of Clinical Microbiology*, **27**, 2218–21.

Kapperud, G., Lassen, J., Lauwers, S. and Rosef, O. (1984) Serotyping and biotyping of *Campylobacter jejuni* and *Campylobacter coli* from sporadic cases and outbreaks in Norway. *Journal of Clinical Microbiology*, **19**, 157–60.

Katouli, M., Kühn, I., Brauner, A. *et al.* (1992) Application of biochemical fingerprinting to the investigation of clonal groups of *Salmonella* of serotype Havana. *Journal of Medical Microbiology*, **36**, 382–8.

Kramer, F.R., Tyagi, S., Guerra, C.E. *et al.* (1991) Q-ß amplification assays, in *Rapid Methods and Automation in Microbiology and Immunology* (eds A. Vaheri, R.C. Tilton and A. Balows), Springer-Verlag, Berlin, pp. 17–22.

Kühn, I., Tullus, K. and Burman, L.G. (1991) The use of the PhP-KE biochemical fingerprinting system in epidemiological studies of faecal *Enterobacter cloacae* strains from infants in Swedish neonatal wards. *Epidemiology and Infection*, **107**, 311–9.

Lammler, C. (1990) Typing of *Actinomyces pyogenes* by its production and susceptibility to bacteriocin-like inhibitors. *Zentralblatt für Bakteriologie International Journal of Medical Microbiology*, **273**, 173–8.

Lányi, B. (1984) Biochemical and serological characterization of *Citrobacter*, in *Methods in Microbiology*, vol. 15 (ed T. Bergan), Academic Press, London, pp. 143–71.

Lányi, B. and Bergan, T. (1978) Serological characterization of *Pseudomonas aeruginosa*, in *Methods in Microbiology*, vol. 10 (eds T. Bergan and J.R. Norris), Academic Press, London, pp. 93–168.

Larsson, P. (1984) Serology of *Proteus mirabilis* and *Proteus vulgaris*, in *Methods in Microbiology*, vol. 14 (ed T. Bergan), Academic Press, London, pp. 187–214.

Lawal, S.F. and Dosunmuogunbi, O. (1986) A new scheme for serotyping group-G streptococci. *Journal of Medical Microbiology*, **21**, 275–7.

Li, Z.S., Belanger, M. and Jacques, M. (1991) Serotyping of Canadian isolates of *Treponema hyodysenteriae* and description of 2 new serotypes. *Journal of Clinical Microbiology*, **29**, 2794–7.

Lindeberg, A.A. and Le Minor, L. (1984) Serology of *Salmonella*, in *Methods in Microbiology*, vol. 15 (ed T. Bergan), Academic Press, London, pp. 1–141.

Magee, J.T., Hindmarch, J.M. and Meechan, D.F. (1983) Identification of staphylococci by pyrolysis gas liquid chromatography. *Journal of Medical Microbiology*, **16**, 483–95.

Magee, J.T., Hindmarch, J.M., Duerden, B.I. and Mackenzie, D.R.W. (1988) Pyrolysis mass spectrometry as a method for inter-strain discrimination of *Candida albicans*. *Journal of General Microbiology*, **134**, 2841–7.

Magee, J.T., Hindmarch, J.M., Bennett, K.W. *et al.* (1989a) A pyrolysis mass spectrometry study of Fusobacteria. *Journal of Medical Microbiology*, **28**, 227–36.

Magee, J.T., Hindmarch, J.M., Burnett, I.A. and Pease, A. (1989b), Epidemiological typing of *Streptococcus pyogenes* by pyrolysis mass spectrometry. *Journal of Medical Microbiology*, **30**, 273–8.

Magot, M., Carlier, P. and Popoff, M.R. (1983) Identification of *Clostridium butyricum* and *Clostridium beijerinckii* by gas liquid chromatography and sugar fermentation – correlation with DNA homologies and electrophoretic patterns. *Journal of General Microbiology*, **129**, 2837–45.

Maliwan, N., Reid, R.W. and Katzen, R.B. (1984) Gas–liquid chromatrography for rapid diagnosis and monitoring of invasive candidal infections and candidemia. *Archives of Pathology and Laboratory Medicine*, **108**, 108–11.

Maliwan, N., Reid, R.W., Pliska, S.R. *et al.* (1988) Identifying *Mycobacterium tuberculosis* cultures by gas–liquid chromatography and a computer-aided pattern-recognition model. *Journal of Clinical Microbiology*, **26**, 182–7.

Marmet, D., Bornstein, N. and Fleurette, J. (1988) Identification of *Legionella* by gas–liquid chromatography (GLC) of fatty acids and high-performance liquid chromatography (HPLC) of ubiquinones. *Annales de Biologie Clinique*, **46**, 371–5.

Maxam, A. M. and Gilbert, W. (1977) A new method for sequencing DNA. *Proceedings of the National Academy of Sciences of the USA*, **74**, 560–4.

McArthur, W.P., Stroup, S. and Wilson, L. (1986) Detection and serotyping of *Actinobacillus actinomycetemcomitans* isolates on nitrocellulose paper blots with monoclonal antibodies. *Journal of Clinical Periodontology*, **13**, 684–91.

McBridge, J.S., Welsby, P.D. and Walliker, D. (1984) Serotyping *Plasmodium falciparum* from acute human infections using monoclonal antibodies. *Transactions of the Royal Society of Tropical Medicine and Hygiene*, **78**, 32–4.

McCullough, N.B. (1978) Identification of the species and biotypes within the genus *Brucella*, in *Methods in Microbiology*, vol. 10 (eds T. Bergan and J.R. Norris), Academic Press, London, pp. 201–25.

McLauchlin, J., Audurier, A. and Taylor, A.G. (1986) Aspects of the epidemiology of human *Listeria monocytogenes* infections in Britain 1967–1984 – the use of serotyping and phage typing. *Journal of Medical Microbiology*, **22**, 367–77.

Meitert, T. and Meitert, E. (1978) Usefulness, applications and limitations of epidemiological typing methods to elucidate nosocomial infections and the spread of communicable diseases, in *Methods in Microbiology*, vol. 10, (eds T. Bergan and J.R. Norris), Academic Press, London, pp. 1–37.

Milch, H. (1978) Phage typing of *Escherichia coli*, in *Methods in Microbiology*, vol. 11 (eds T. Bergan and J.R. Norris), Academic Press, London, pp. 88–155.

Mills, S.D., Congi, R.V., Hennessy, J.N. and Penner, J.L. (1991) Evaluation of a simplified procedure for serotyping *Campylobacter jejuni* and *Campylobacter coli* which is based on the O-antigen. *Journal of Clinical Microbiology*, **29**, 2093–8.

Moore, L.J. and Lepper, A.W.D. (1991) A unified typing scheme for *Moraxella bovis*. *Veterinary Microbiology*, **29**, 75–83.

Moyes, A. and Young, H. (1991) Epidemiologic typing of *Neisseria gonorrhoeae* – a comparative analysis of 3 monoclonal antibody serotyping panels. *European Journal of Epidemiology*, **7**, 311–9.

Mukerjee, S. (1978) Principles and practice of typing *Vibrio cholerae*, in *Methods in Microbiology*, vol. 12, (eds T. Bergan and J.R. Norris), Academic Press, London, pp. 52–115.

Mullis, K.B. and Faloona, F.A. (1987) Specific synthesis of DNA in vitro via a polymerase-catalyzed chain reaction, in *Methods in Enzymology*, vol. 155 (eds R. Wu, L. Grossman and K. Moldave), Academic Press, New York, pp. 335–50.

Murray, R.G.E., Brenner, D.J., Colwell, R.R. *et al.* (1990) Report of the ad hoc committee on approaches to taxonomy within the Proteobacteria. *International Journal of Systematic Bacteriology*, **40**, 213–5.

Nawa, M., Ichikawa, Y. and Inouye, S. (1985) Serotyping of Dengue viruses by an enzyme-linked immunosorbent assay. *Japanese Journal of Medical Science and Biology*, **38**, 217–21.

Norrung, V., Munch, B. and Larsen, H.E. (1987) Occurrence, isolation and serotyping of *Erysipelothrix rhusiopathiae* in cattle and pig slurry. *Acta Veterinaria Scandinavica*, **28**, 9–14.

Orr, K., Gould, F.K., Sisson, P.R. *et al.* (1991) Rapid inter-strain comparison by pyrolysis mass spectrometry in nosocomial infection with *Xanthomonas maltophilia*. *Journal of Hospital Infection*, **17**, 187–95.

Ørskov, F. and Ørskov, I. (1984a) Serotyping of *Escherichia coli*, in *Methods in Microbiology*, vol. 14 (ed T. Bergan), Academic Press, London, pp. 43–112.

Ørskov, I. and Ørskov, F. (1984b) Serotyping of *Klebsiella*, in *Methods in Microbiology*, vol. 14 (ed T. Bergan), Academic Press, London, pp. 143–64.

Ota, F., Kiso, M., Fukui, K. *et al.* (1984) Serotyping of *Streptococcus mutans* with agglutinating antibodies. *Journal of Dental Research*, **63**, 481.

Parker, M.T. (1972) Phage typing of *Staphylococcus aureus*, in *Methods in Microbiology*, vol. 7B (eds J.R. Norris and D.W. Ribbons), Academic Press, London, pp. 1–28.

Pasaribu, F.H., Lammler, C. and Blobel, H. (1985) Serotyping of bovine and human group B streptococci by coagglutination. *IRCS Medical Science-Biochemistry*, **13**, 24–5.

Pinner, M.S. and Markham, P.G. (1990) Serotyping and strain identification of maize streak virus isolates. *Journal of General Virology*, **71**, 1635–40.

Polonelli, L., Castagnola, M. and Morace, G. (1986) Identification and serotyping of *Microsporum canis* isolates by monoclonal antibodies. *Journal of Clinical Microbiology*, **23**, 609–15.

Ragland, N. and Tagg, J. (1990) Applications of bacteriocin-like inhibitory substance (BLIS) typing in a longitudinal study of the oral carriage of beta-haemolytic streptococci by a group of Dunedin schoolchildren. *Zentralblatt für Bakteriologie International Journal of Medical Microbiology*, **274**, 101–8.

Redmond, W.B., Bates, J.H. and Engel, H.W. (1979) Methods of bacteriophage typing of mycobacteria, in *Methods in Microbiology*, vol. 13 (eds T. Bergan and J.R. Norris), Academic Press, London, pp. 345–76.

Robinson, A., Ashworth, L.A.E. and Irons, L.I. (1989) Serotyping *Bordetella pertussis* strains. *Vaccine*, **7**, 491–4.

Sanchis, R., Polveroni, G. and Abadie, G. (1988) Serotyping of *Pasteurella haemolytica* – comparison and adjustment of antigenic extracts and techniques. *Revue de Medecine Veterinaire*, **139**, 719–22.

Sanger, F., Nicklon, S. and Coulson, A.R. (1977) DNA sequencing with chain-terminating inhibitors. *Proceedings of the National Academy of Sciences of the USA*, **74**, 5463–7.

Saragea, A., Maximescu, P. and Meitert, E. (1979) *Corynebacterium diphtheriae*. Microbiological methods used in clinical and epidemiological investigations, in *Methods in Microbiology*, vol. 13 (eds T. Bergan and J.R. Norris), Academic Press, London, pp. 62–176.

Schwartz, D.C. and Cantor, C.R. (1984) Separation of yeast chromosome-sized DNAs by pulsed field gradient gel electrophoresis. *Cell*, **37**, 67–75.

Senior, B.W. (1987) The typing of *Morganella morgani* by bacteriocin production and sensitivity. *Journal of Medical Microbiology*, **23**, 33–9.

Shimada, T., Kosako, Y., Inoue, K. *et al.* (1991) *Vibrio fluvialis* and *V. furnissii* serotyping scheme for international use. *Current Microbiology*, **22**, 335–7.

Sisson, P.R., Freeman, R., Gould, F.K. and Lightfoot, N.F. (1991a) Strain differentiation of nosocomial isolates of *Pseudomonas aeruginosa* by pyrolysis mass spectrometry. *Journal of Hospital Infection*, **19**, 137–40.

Sisson, P.R., Freeman, R., Lightfoot, N.F. and Richardson, I.R. (1991b) Incrimination of an environmental source of a case of Legionnaires' disease by pyrolysis mass spectrometry. *Epidemiology and Infection*, **107**, 127–32.

Sisson, P.R., Freeman, R., Magee, J.G. and Lightfoot, N.F. (1991c) Differentiation between mycobacteria of the *Mycobacterium tuberculosis* complex by pyrolysis mass spectrometry. *Tubercle*, **72**, 206–9.

Skalka, B. (1987) Bacteriocin typing of staphylococci. *Folia Microbiologica*, **32**, 516.

Slade, M.B. and Tiffin, A.I. (1984) Biochemical and serological characterization of *Erwinia*, in *Methods in Microbiology*, vol. 15 (ed T. Bergan), Academic Press, London, pp. 227–93.

Slopek, S. (1978) Phage typing of *Klebsiella*, in *Methods in Microbiology*, vol. 11 (eds T. Bergan and J.R. Norris), Academic Press, London, pp. 193–222.

Smyth, C.J., Matthews, H., Halpenny, M.K. *et al.* (1987) Biotyping, serotyping and phage typing of *Streptococcus faecalis* isolated from dental plaque in the human mouth. *Journal of Medical Microbiology*, **23**, 45–54.

Sorensen, U.B.S. and Larsen, J.L. (1986) Serotyping of *Vibrio anguillarum*. *Applied and Environmental Microbiology*, **51**, 593–7.

Stead, D.E. (1992) Grouping of plant-pathogenic and some other *Pseudomonas* spp. by using cellular fatty-acid profiles. *International Journal of Systematic Bacteriology*, **42**, 281–95.

Stoakes, L., Kelly, T., Schieven, B. *et al.* (1991) Gas–liquid chromatographic analysis of cellular fatty acids for identification of gram-negative anaerobic bacilli. *Journal of Clinical Microbiology*, **29**, 2636–8.

Stoev, I. (1979) Methods of typing *Erysipelothrix insidiosa*, in *Methods in Microbiology*, vol. 13 (eds T. Bergan and J.R. Norris), Academic Press, London, pp. 51–60.

Stringer, J. (1984) Phage typing of *Streptococcus agalactiae*, in *Methods in Microbiology*, vol. 16 (ed T. Bergan), Academic Press, London, pp. 1–22.

Sutter, V.L., Citron, D.M., Edelstein, M.A.C. and Finegold, S.M. (1985) *Wadsworth Anaerobic Bacteriology Manual*, 4th edn, Star Publishing, Belmont.

Tagg, J.R. and Martin, D.R. (1984) Evaluation of a typing scheme for group A streptococci based upon bacteriocin-like inhibitor production. *Zentralblatt für Bakteriologie Mikrobiologie und Hygiene Series A*, **257**, 60–7.

Tamura, K., Sakazaki, R., McWhorter, A.C. and Kosako, Y. (1988) *Edwardsiella tarda* serotyping scheme for international use. *Journal of Clinical Microbiology*, **26**, 2343–6.

Tompkins, L.S., Troup, N., Labigne-Roussel, A. and Cohen, M.L. (1986) cloned, random chromosomal sequences as probes to identify *Salmonella* species. *Journal of Infectious Diseases*, **154**, 156–62.

Toma, S., Lesiak, G., Magus, M. *et al.* (1988) Serotyping of *Clostridium difficile*. *Journal of Clinical Microbiology*, **26**, 426–8.

Traub, W.H. (1978) Bacteriocin typing of clinical isolates of *Serratia marcescens*, in *Methods in Microbiology*, vol. 11 (eds T. Bergan and J.R. Norris), Academic Press, London, pp. 223–42.

Traub, W.H. (1989) *Acinetobacter baumannii* serotyping for delineation of outbreaks of nosocomial infection. *Journal of Clinical Microbiology*, **27**, 2713–6.

Traub, W.H. (1990) Serotyping of clinical isolates of *Acinetobacter* serovars of genospecies–3. *Zentralblatt für Bakteriologie International Journal of Medical Microbiology*, **273**, 12–23.

Traub, W.H. (1991) Comparative biotyping, bacteriocin typing, and serogrouping (O-antigens) of *Serratia liquefaciens*. *Zentralblatt für Bakteriologie International Journal of Medical Microbiology*, **275**, 200–10.

Van Brunt, J. (1990) Amplifying genes: PCR and its alternatives. *Biotechnology*, **8**, 291–4.

Verschraegen, G., Claeys, G., Delanghe, M. and Pattyn, P. (1988) Serotyping and phage typing to identify *Enterobacter cloacae* contaminating total parenteral nutrition. *European Journal of Clinical Microbiology and Infectious Diseases*, **7**, 306–7.

Waltman, W.D., McDaniel, L.S., Andersson, B. *et al*, (1988) Protein serotyping of *Streptococcus pneumoniae* based on reactivity to 6 monoclonal antibodies. *Microbial Pathogenesis*, **5**, 159–67.

Wang, S.P. and Grayston, J.T. (1991) Serotyping of *Chlamydia trachomatis* by indirect fluorescent-antibody staining of inclusions in cell culture with monoclonal–antibodies. *Journal of Clinical Microbiology*, **29**, 1295–8.

Watson, G.N. (1985) The assessment and application of a bacteriocin typing scheme for *Clostridium perfringens*. *Journal of Hygiene*, **94**, 69–79.

Wayne, L.G., Brenner, D.J., Colwell, R.R. *et al.* (1987) Report of the ad hoc committee on reconciliation of approaches to bacterial systematics. *International Journal of Systematic Bacteriology*, **37**, 463–4.

Wedege, E., Hoiby, E.A., Rosenqvist, E. and Froholm, L.O. (1990) Serotyping and subtyping of *Neisseria meningitidis* isolates by coagglutination, dot-blotting and ELISA. *Journal of Medical Microbiology*, **31**, 195–201.

Analysis of nucleic acid profiles 2

2.1 INTRODUCTION

As outlined in Chapter 1, analysis of the genetic material of a microorganism offers the major advantage that it is not subject to phenotypic variation since it does not rely on the expression of particular properties belonging to the organism. In most microorganisms, the genetic material consists of double-stranded DNA (dsDNA), with the specific sequence of nucleotide base pairs providing the information content.

The major genetic structures in cells are chromosomes; in eukaryotic organisms these may be complicated structures that also involve proteins, but in the vast majority of bacteria so far examined, the chromosome comprises a single large circular dsDNA molecule. The two possible known exceptions to this rule are *Rhodobacter sphaeroides*, which may have two distinct circular chromosomes (Suwanto and Kaplan, 1989), and the spirochaete *Borrelia burgdorferi*, which seems to have a linear chromosome (Ferdows and Barbour, 1989).

Plasmids are autonomous genetic elements that can replicate independently of the main chromosome and can be separated physically from it in the laboratory. Plasmids often encode useful additional properties for the cell, such as antibiotic resistance and metabolic activities (reviewed by Stanisich, 1988). Bacterial plasmids are generally circular dsDNA molecules that are covalently closed in each strand. The main exceptions to this rule are the so-called 'linear' plasmids found in *Borrelia* spp. (Barbour and Garon, 1987) and *Streptomyces* spp. (Kinashi, Shimaji and Sakai, 1987). These linear plasmids seem to have covalently closed ends in a structure similar to that found in some viruses (Barbour and Garon, 1987). Plasmids have been found in the vast majority of bacterial genera in which they have been sought (Stanisich, 1988), and have also been reported in some lower eukaryotic microorganisms such as yeasts (Broach, 1981).

This chapter is concerned with methods that examine the size and structure of individual nucleic acid molecules, either plasmids or chromosomes. Such methods cannot be used for taxonomic identification of a particular microorganism, but are being used increasingly by more adventurous diagnostic laboratories for typing purposes.

2.2 PLASMID ANALYSIS

2.2.1 PLASMID PROFILES

(a) Isolation of plasmid DNA

The plasmid content of a cell normally comprises less than 5% of the total DNA, and examination of plasmid profiles first requires the separation of plasmid DNA from chromosomal DNA. Normal cell lysis procedures invariably result in random shearing of the large chromosomal molecules to generate linear fragments. Early plasmid isolation procedures relied on lengthy and complicated methods that involved the separation of covalently closed circles of plasmid DNA from chromosomal DNA fragments by ultracentrifugation on caesium chloride gradients containing high concentrations of ethidium bromide. While such techniques are still used occasionally in molecular biology research laboratories for preparing large quantities of pure plasmid DNA, there are now numerous rapid 'mini-prep' methods for isolating plasmids on a small scale, and it is these rapid methods which can be applied to large numbers of bacterial isolates for typing purposes and epidemiological

Table 2.1 Examples of rapid 'mini-prep' methods for isolation of plasmid DNA

Gram-negative bacteria:	Birnboim and Doly (1979) Klein *et al.* (1980) Holmes and Quigley (1981) Ish-Horowicz and Burke (1981) Kado and Liu (1981) Schaberg *et al.* (1981) Takahashi and Nagano (1984) Bennett *et al.* (1986) Olsen (1990) Domenico *et al.* (1992)
Gram-positive bacteria:	Wilson *et al.* (1978) Dunkle and Sippel (1984) Friedrich and Lütticken (1984) Klaenhammer *et al.* (1984) Nahaie *et al.* (1984) Takahashi and Nagano (1984) Roberts *et al.* (1986) Somkuti and Steinberg (1986)

studies. However, Gram-negative and Gram-positive bacteria require slightly different approaches, and different laboratories have their own preferences from amongst the various methods available (Table 2.1).

To isolate plasmid DNA, the bacterial cell has first to be lysed, followed by removal of the bulk of the chromosomal DNA. This can be done either by gentle lysis followed by a 'clearing' centrifugation step, or by complete lysis followed by removal of the chromosomal DNA after selective precipitation. Although the individual protocols differ in their precise details, the most popular approach (in summary) involves first weakening the bacterial cell wall, usually either by boiling or by treatment with lysozyme, followed by complete lysis in the presence of alkali by the addition of EDTA and a detergent such as sodium dodecyl sulphate (SDS) or Triton X–100. High-molecular-mass genomic DNA and protein is then precipitated selectively following neutralization with acetate, and is then removed by centrifugation to leave the plasmid DNA in solution. Following transfer to a fresh centrifuge tube, plasmid DNA is then, in turn, precipitated by the addition of alcohol, recovered by centrifugation, and re-dissolved in a desired volume of buffer.

Such methods are fast, reproducible, and provide several micrograms of plasmid DNA which can be digested readily by appropriate restriction endonucleases (section 2.2.2). A single worker could reasonably expect to perform at least 100 such preparations in the course of one day. The preparations can be performed entirely in 1.5 ml disposable Eppendorf tubes, and the only major piece of apparatus required is a small, relatively inexpensive, bench-top centrifuge capable of spinning such tubes at 12 000–15 000 *g*, together with a simple electrophoresis apparatus and ultraviolet light source for subsequent visualization of the plasmid DNA (see below). The best yields are obtained with smaller plasmids, partly because of their larger copy numbers and partly because they are less prone to physical damage, but plasmids of >100 kb in size can be visualized if appropriate care in handling is taken. All 'mini-prep' preparations are normally contaminated with substantial amounts of RNA, which co-migrates with, and masks, DNA fragments which are <500 bp in size. This is rarely a problem when examining plasmids, but in any event, the RNA can be removed readily by a brief treatment of the samples with ribonuclease A.

(b) Agarose gel electrophoresis of plasmid DNA

Once plasmid DNA has been isolated, it must be visualized before it can be analysed. Normally this is achieved by electrophoresis through an agarose gel to give an indication of the number and sizes of any plasmids present (the **plasmid profile**). Vertical gel formats were used widely when the technique was first introduced, but these have now diminished greatly in popularity. For ease of handling, most plasmid studies are now performed with horizontal slab gels and an agarose concentration which varies according to the anticipated sizes of the plasmids to be visualized (the larger the plasmids, the lower the percentage of agarose which is needed to resolve them). With horizontal gels it is possible to use agarose concentrations as low as 0.3%, but at such a concentration the gel becomes very sloppy and is difficult to handle. Most determinations of plasmid profiles therefore use an agarose concentration of either 0.7% or 1%.

Almost any buffer system could in theory be used for electrophoresis of plasmid DNA, but those listed in Table 2.2 are used most widely. Concentrated stock solutions (normally ×10) are prepared and diluted as required. Horizontal gels are just covered by a layer of buffer during the electrophoresis run. This is termed **'submarine' gel electrophoresis**. The electrical resistance of the gel is very similar to that of the buffer so that, with a thin layer of buffer, most of the current passes through the gel itself. Deeper immersion of the gel results in the electrophoresis process becoming less efficient. The submarine technique has the major advantage that the gel is prevented from drying out during the course of the experiment.

Table 2.2 Common electrophoresis buffer systems

Tris-borate (TBE)	89 mM Tris-borate 89 mM Boric acid 2 mM EDTA pH8.0
Tris-acetate (TAE)	40 mM Tris-acetate 1 mM EDTA pH8.0
Tris-phosphate (TPE)	80 mM Tris-phosphate 2 mM EDTA pH8.0

Gels can be of almost any size, and many different inexpensive formats are available commercially, but to optimize resolution of large plasmids it is desirable to perform electrophoresis for as long as possible on as large a gel as possible. Overnight separations with voltages of 2 V/cm and gels of about 50 × 20 cm can be used for large plasmids (>50 kb), but small plasmids (<15 kb) can be separated adequately on small gels (10 × 10 cm) in approximately 1 h with voltages of 5 V/cm. Running a small gel for extended periods will result in smaller plasmids migrating off the end of the gel.

Once electrophoretic separation is complete, the various plasmids present are visualized by staining with ethidium bromide and viewing

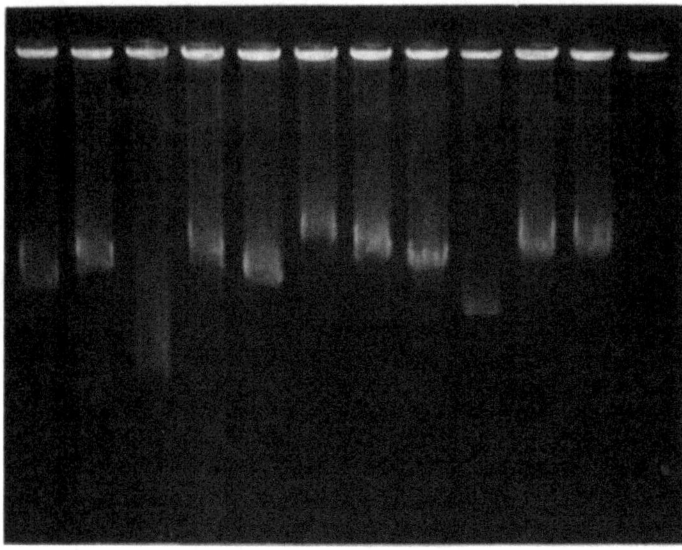

Figure 2.1 Example of plasmid profile analysis of *Eschericha coli* strains, each carrying a single plasmid of different molecular size.

in ultraviolet light (c. 310 nm), ideally on a UV transilluminator which illuminates the gel from underneath. Ethidium bromide fluoresces strongly when bound to DNA, so that plasmid DNA molecules appear as orange bands in the gel. Some authors recommend incorporation of ethidium bromide into both the gel and the running buffer before electrophoresis, but personal experience suggests that postelectrophoresis staining with a solution of ethidium bromide dissolved in electrophoresis buffer (5 mg/l) for a short period (10 min), followed by a rinse with distilled water, results in lower background fluorescence. Finally, the plasmid profiles of the strains being examined should be recorded by photography through a red filter (Polaroid type 667 instant film is the most convenient and widely used system). Figure 2.1 shows an example of straightforward plasmid profile analysis in which each bacterial strain examined carries a plasmid of different molecular size.

(c) Problems in analysis of plasmid profiles

Several potential problems can be encountered when interpreting plasmid profiles seen in agarose gels. First, a single plasmid can generate at least three different bands, corresponding to three different molecular forms (Figure 2.2). Initially, a plasmid preparation will consist mostly of covalently closed circular (CCC) DNA molecules, but a small amount

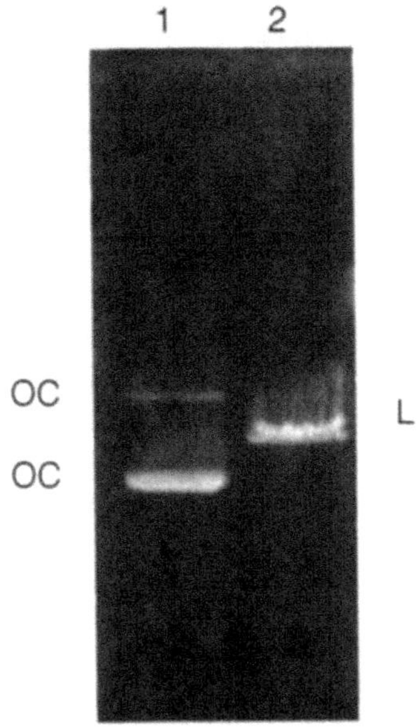

Figure 2.2 Different molecular forms of plasmid pBR322. Track 1, freshly-isolated plasmid DNA showing covalently closed circular (CC) and 'nicked' open circular (OC) forms; track 2, the same preparation following conversion to the linear (L) form by digestion with restriction endonuclease *Eco*RI.

of the 'nicked' open circular (OC) form of the same DNA molecules will often also be present. The proportion of OC DNA is likely to be higher in plasmid preparations that have been stored for any length of time; this is particularly the case for preparations which contain larger plasmids. In addition, linear plasmid molecules may also be visible. Each type of DNA molecule has a different electrophoretic mobility which may, in turn, be affected by alterations in voltage, running buffer or the agarose concentration in the gel. Thus, even though usually the CCC form migrates the fastest, followed by the linear and OC forms, the relative gel positions may change under certain electrophoretic conditions (Kieser, 1984a).

Other complications arise from the fact that some plasmid preparations, particularly those of small plasmids, may contain a mixture of oligomers consisting of integral multiples of the expected monomer. Meanwhile, contaminating chromosomal DNA fragments appear usually as a somewhat diffuse band, in a position that depends on the plasmid preparation method, while a 'smear' right down a sample track means that the preparation has either been grossly overloaded or is contaminated with fragments of random size. Contaminating RNA appears as a large diffuse area of fluorescence which normally migrates well in front of all but the very smallest plasmids.

All of these factors mean that care must be taken in deciding on the identity of DNA bands visualized on a gel, particularly when there is more than one plasmid present. So far as plasmid analysis for typing purposes is concerned, there may be confusion as to whether a particular band is the CCC form of one plasmid or the OC form of another. Although various methods have been described which convert CCC DNA to its OC form before electrophoresis (e.g. Hintermann *et al.*, 1981), these add to the time required to complete the analysis and, from a typing point of view, it may be simpler just to ensure that all samples are prepared in exactly the same way and visualized on a single gel.

The only way of comparing plasmid profiles obtained on different gels is to determine the individual molecular sizes of each of the different plasmids. The size of unknown plasmid molecules, expressed in either megadaltons (Mdal) or kilobase pairs (kb), is determined by plotting their mobility against a standard curve obtained from the mobilities of a set of standard plasmid molecules, run on the same gel, of known molecular size. A common mistake is to attempt to determine the sizes of circular plasmid molecules by comparison with linear DNA fragments obtained by restriction digestion (section 2.2.2). The differing mobilities of the different plasmid forms (CCC, OC and linear) means that CCC molecules can be sized only by comparison with other CCC molecules of known size. This can sometimes be difficult if it is not clear whether a particular plasmid band is the CCC or OC form.

(d) Applications of plasmid profile analysis for typing purposes

Plasmid analysis cannot, of course, be used for typing purposes if the strains being examined lack plasmids. It is also worth noting at the outset that it is not always true that identical plasmid profiles are indicative of an epidemiological relationship, since it is quite possible that evolutionally-related organisms will have acquired a similar basic collection of plasmids. Indeed, studies involving salmonellae have shown that particular plasmid patterns are broadly stable among isolates from cases of infection occurring in widely separated geographical locations over a period of several years (Taylor *et al.*, 1982; Riley *et al.*, 1983a; Threlfall, Hall and Rowe, 1986). Other important considerations include the fact that divergence within plasmid fingerprints can occur very rapidly in the absence of selective pressure while, in contrast, serotype-associated plasmids are extremely stable and cannot be used for typing on an epidemiological basis. Providing the above limitations are taken into account, plasmid analysis can be applied readily to the investigation and typing of many different microorganisms (Table 2.3).

Table 2.3 Examples of microorganisms investigated by plasmid profile analysis

Organism	Reference
Acinetobacter spp.	Bouvet (1991)
Aeromonas hydrophila	Noterdaeme *et al.* (1991)
Bacillus cereus	Agata *et al.* (1990)
Borrelia burgdorferi	Schwan *et al.* (1988)
Campylobacter spp.	Lastovica and Ambrosio (1986)
Clavibacter michiganensis	Finnen *et al.* (1990)
Clostridium difficile	Steinberg *et al.* (1987)
Edwardsiella ictaluri	Speyerer and Boyle (1987)
Enterobacter cloacae	Wang *et al.* (1991)
Escherichia coli	Scotland *et al.* (1989)
Haemophilus ducreyi	Abeck *et al.* (1988)
Haemophilus pleuropneumoniae	Huether *et al.* (1987)
Helicobacter pylori	Dworkin *et al.* (1991)
Lactobacillus plantarum	von Husby and Nes (1986)
Legionella pneumophila	Mayer *et al.* (1983)
Listeria spp.	Fistrovici *et al.* (1990)
Moraxella bovis	McDonald and Pugh (1986)
Mycobacterium spp.	Meissner and Falkinham (1986)
Neisseria gonorrhoeae	Mee *et al.* (1988)
Pasteurella haemolytica	Boyce and Morter (1986)
Proteus spp.	Hawkey *et al.* (1986)
Pseudomonas aeruginosa	Poh *et al.* (1988)
Pseudomonas cepacia	John and Twitty (1986)
Rhizobium meliloti	Hartmann and Amarger (1991)
Salmonella spp.	Threlfall *et al.* (1986)
Shigella dysenteriae	Gebre-Yohannes and Drasar (1990)
Staphylococcus aureus	Dodd *et al.* (1988)
Staphylococcus epidermidis	Hartstein *et al.* (1987)
Streptococci	Davies *et al.* (1981)
Streptomycetes	Hotta *et al.* (1980)
Thiobacillus ferrooxidans	Valenti *et al.* (1990)
Vibrio salmonicida	Wiik *et al.* (1989)
Yersinia enterocolitica	Bottone *et al.* (1985)
Yersinia ruckeri	Degrandis and Stevenson (1982)
Zymomonas mobilis	Yablonsky *et al.* (1988)

The method works best with isolates that contain multiple plasmids, and plasmid analysis has therefore been applied extensively to staphylococci, particularly *Staphylococcus epidermidis* (Parisi, 1985; Schaberg and Zervos, 1986). It should, however, be noted that while many isolates of *S. epidermidis* contain multiple plasmids when first isolated, some of these plasmids seem to be lost on prolonged storage (Parisi, 1985). Other important applications have included numerous investigations of outbreaks of salmonellosis and shigellosis (e.g. Threlfall *et al.*, 1986; Gebre-Yohannes and Drasar, 1990), and many studies have shown that plasmid analysis can be applied successfully to the investigation of episodes of nosocomial infection by bacteria for which no known pre-existing typing scheme exists.

2.2.2 PLASMID RESTRICTION ENDONUCLEASE FINGERPRINTS

As discussed in the previous section, problems of interpretation may arise if different molecular forms (CCC, OC, or linear) of a single plasmid are visualized on the same gel, thereby giving three bands where only one plasmid is present. In addition, problems may arise when using plasmid analysis for typing purposes, particularly if preparations contain large plasmids, because agarose gel electrophoresis is not a sensitive method of detecting small differences between plasmids of similar size. Similarly, simple analysis of plasmid profiles will not detect the difference between two dissimilar plasmids of identical size. Such problems can be overcome by using enzymes termed **restriction endonucleases** to generate '**plasmid fingerprints**'. Cutting plasmid DNA with restriction endonucleases adds only about 1 h to the procedure and provides much additional typing information. This technique is described more fully in the following sections.

(a) Properties of restriction endonucleases

Restriction endonucleases are enzymes that recognize a specific base sequence of DNA, typically 4–6 base pairs (bp) long, and then cleave the DNA at a defined position in relation to the specific recognition sequence. Over 400 restriction endonucleases have been isolated from bacteria and characterized (Roberts and Macelis, 1991). Some of the restriction endonucleases used most commonly for plasmid fingerprinting are listed in Table 2.4, together with their respective recognition sequences. All of these enzymes are named after the bacterial species from which they were first isolated, with the first letter of their designation derived from the genomic name, and the second two letters derived from the specific name. Restriction endonucleases isolated from different species of bacteria may have the same recognition sequence; these are termed **isoschizomers**. The natural function of restriction endonucleases is thought to be the protection of bacteria from foreign DNA. Methylation of DNA at the recognition sequence prevents cleavage. Foreign DNA lacks this methylation and is therefore cleaved (Old and Primrose, 1985).

The specificity of DNA cleavage by restriction endonucleases means that complete digestion of a particular sequence of DNA by a specified enzyme or combination of enzymes will result in the production of a reproducible set of linear fragments, generated according to the frequency and location of the specific enzyme recognition sequence(s).

Table 2.4 Common restriction endonucleases used for the generation of DNA fingerprints

Enzyme	Recognition sequence*
*Apa*I	GGGCC/C
*Bam*HI	G/GATCC
*Bgl*I	GCCNNNN/NGGC
*Bgl*II	A/GATCT
*Cla*I	AT/CGAT
*Eco*RI	G/AATTC
*Eco*RV	GAT/ATC
*Hin*dIII	A/AGCTT
*Hin*fI	G/ANTC
*Hpa*II	C/CGG
*Kpn*I	GGTAC/C
*Mlu*I	A/CGCGT
*Pst*I	CTGCA/G
*Pvu*II	CAG/CTG
*Sac*I	GAGCT/C
*Sca*I	AGT/ACT
*Sma*I	CCC/GGG
*Stu*I	AGG/CCT
*Xho*I	C/TCGAG

* N indicates a random nucleotide base; / indicates cutting site in relation to the recognition sequence.

Plasmid DNA molecules can, therefore, be compared by examining the number and size of fragments generated by digestion of the DNA with restriction endonucleases. The pattern of fragments produced is termed the **plasmid fingerprint**, while variations observed between related molecules are termed **restriction fragment length polymorphisms (RFLPs)**.

(b) Generation of plasmid fingerprints

The first step in the generation of a plasmid fingerprint is the preparation of a reasonably pure sample of plasmid DNA for use as a substrate for restriction endonuclease digestion. This can be accomplished readily by most of the rapid methods outlined in section 2.2.1. Depending on the plasmid DNA molecules to be examined, one or more restriction endonucleases are then selected on the basis of how many fragments they are likely to produce. The number of fragments should be sufficient to enable good discrimination, but not so many that they overlap or are coincidental on the gel, thereby resulting in a crowded fingerprint which is difficult to interpret.

The problem of enzyme selection is usually more important for digestion of chromosomal DNA (section 2.3) than it is for plasmid DNA,

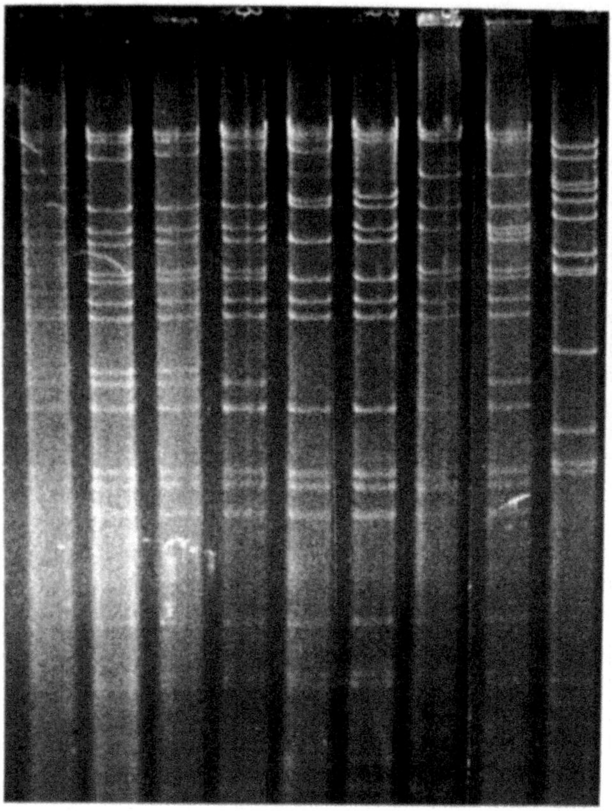

Figure 2.3 Example of plasmid fingerprints generated by *Eco*RI digestion of plasmids with similar molecular sizes.

when trial and error will usually suffice. However, in general, restriction enzymes which have 4 bp recognition sites will cut DNA more frequently than those which recognize 6 bp sequences.

Most commercially available restriction endonucleases are now supplied by their manufacturer complete with detailed instructions and an appropriate buffer concentrate. This enables digestions to be performed reproducibly, rapidly and at high efficiency. Following digestion of plasmid DNA, normally for a period of about 1 h, the resulting DNA fragments are visualized by ethidium bromide staining following separation by electrophoresis through an agarose or polyacrylamide gel. Agarose concentrations required to give optimum separation of linear DNA fragments over different size ranges are listed in Table 2.5, while Figure 2.3 shows examples of the plasmid fingerprints generated from plasmids of similar molecular size following restriction endonuclease digestion.

Table 2.5 Agarose concentrations required for optimum separation of linear DNA fragments in different size ranges

Agarose conc. (%)	Linear size range (kb)
0.3	1.0–70.0
0.5	0.7–45.0
0.8	0.4–20.0
1.0	0.3–10.0
1.2	0.2–8.0
1.5	0.2–6.0
2.0	0.1–5.0

(c) Analysis of plasmid fingerprints

Determination of DNA fragment sizes is not required when plasmid fingerprints are compared side-by-side, but it is difficult to arrange this when large numbers of strains are being examined. In addition, if the precise sizes of the DNA fragments can be recorded, it is then possible to compare fingerprints obtained across time from many different laboratories. Unfortunately, most published data have been obtained from comparisons of migration distances on agarose gels with only a few fragment size standards, followed by somewhat uncritical semilogarithmic interpolation of fragment sizes. Considerable size errors can thereby be generated, which can become a problem when attempting to decide on the similarity of two DNA fragments reported from different laboratories.

Grimont and Grimont (1991) have suggested that the ideal mixture of known standard fragments for determining sizes should consist of at least ten fragments, including fragments that are bigger and smaller than any fragment generated from the plasmid DNA being tested, with the proviso that fragments greater than 20 kb in size may not migrate properly on normal concentration agarose gels. It is also important that the sizes of the standard fragments should have been determined originally by sequencing, rather than by poor interpolation from another set of standards, but this information is not always available for larger standard fragments. The best solution is probably to use one of the various well characterized size standard mixtures which are now available commercially.

Assuming that a suitable mixture of standard size fragments is available, it is then necessary to make precise measurements of migration distances on agarose gels, and ensure that these are not spoilt by subsequent inaccurate calculation of fragment sizes. Various mathematical linear models have been devised that allow the molecular sizes of unknown fragments to be calculated. These include the logarithm of the molecular size versus mobility (Aaij and Borst, 1972; Duggleby, Kinns and Rood, 1981), logarithm of the molecular size versus logarithm of mobility (Meyers *et al.*, 1976), and molecular size versus reciprocal of mobility (Southern, 1979; Elder and Southern, 1983). The last of these three models seems to be one of the most accurate and useful, particularly when it is implemented by use of a least squares analysis (Schaffer and Sederoff, 1981) and adapted to run on personal computers. Other more complex mathematical models are available (Rochelle *et al.*, 1985; Plikaytis *et al.*, 1986), and a comparison of some of the above methods has been published by Owen and Beck (1987).

Manual analysis of complex fingerprint patterns is difficult when large numbers of samples have to be compared. Unfortunately, most conventional densitometers cannot be used for direct screening of fluorescing DNA bands stained with ethidium bromide. However, providing even illumination of the whole gel can be ensured, band patterns can be recorded and quantified by photographing them and then using a densitometer to scan the photographic negatives. Quantification is achieved by simultaneously running and scanning a standard mixture containing known amounts of DNA on the same gel as the unknown samples (Prunell *et al.*, 1977; Goldman and Merrill, 1982). Once the quantitative raw data has been obtained, there are many ways in which it can be handled (e.g. Kieser, 1984b), ranging from fully computerized analysis to simple manual estimation of peak areas.

A possible alternative to photography is the electronic imaging of fluorescent band patterns with a modified television camera equipped with a charge-coupled device (CCD) detector (Sutherland *et al.*, 1987a; Chen and Sutherland, 1989). Such a system provides the image in digital form for immediate analysis by computer, and also has the advantages of a greater dynamic range and a linear response. Several image analysis systems of this type are either available or in the process of development.

(d) Applications of plasmid fingerprinting for typing purposes

Straightforward plasmid profile analysis is readily applicable to isolates that contain multiple plasmids, and restriction endonuclease digestion of plasmid DNA is usually only necessary on rare occasions for typing purposes. The generation of plasmid fingerprints is of more value for epidemiological purposes when only apparently similar single plasmids of relatively large size can be identified, or when similar plasmid profiles occur in strains that cannot be differentiated by other methods. Plasmid fingerprints have been used most extensively to investigate strains of Enterobacteriaceae, and in two landmark investigations of food poisoning outbreaks associated with hamburger consumption in the USA, combined use of plasmid profiles and restriction endonuclease plasmid fingerprint patterns proved to be effective in linking patient isolates with hamburger isolates of *Escherichia coli* 0157:H7 (Riley *et al.*, 1983b; Wells *et al.*, 1983). Table 2.6 lists examples of other organisms with which plasmid fingerprinting has proved to be useful for typing and epidemiological purposes.

Table 2.6 Examples of organisms investigated by restriction endonuclease digestion of plasmid DNA

Organism	Reference
Acinetobacter spp.	Spiliopoulou *et al.* (1992)
Bacteroides fragilis	Beul *et al.* (1985)
Clostridium difficile	Mulligan *et al.* (1988)
Enterobacter cloacae	Wang *et al.* (1991)
Escherichia coli	Wells *et al.* (1983)
Haemophilus influenzae	Loos *et al.* (1989)
Helicobacter pylori	Owen *et al.* (1990)
Lactobacillus spp.	Stahl *et al.* (1990)
Neisseria spp.	Sorensen *et al.* (1985)
Providencia stuartii	Hawkey *et al.* (1984)
Salmonella spp.	Wachsmuth *et al.* (1991)
Serratia spp.	Platt and Somerville (1981)
Staphylococcus aureus	Coia *et al.* (1988)

Plasmids differ considerably in the number of restriction sites which they possess for a given enzyme, and a large number of different enzymes have been used to generate fingerprints. Consequently, it is difficult at present to compare plasmid profiles obtained in different laboratories without employing a wide range of enzymes. Platt *et al.* (1986) suggested that a common strategy should be followed when choosing restriction endonucleases to generate plasmid fingerprints for typing purposes. Such a common strategy would seem to be an excellent idea, but one which does not seem to have gained wide acceptance – most laboratories seem to use whatever enzymes they have to hand!

A mistake which is sometimes made concerns the precise epidemiological or typing significance of two plasmid fingerprints that differ by only one or two fragments. It is important to realise that RFLPs (leading to alterations in the band patterns) can each be generated by a single base pair change in the structure of the plasmid DNA. Such base pair changes can occur spontaneously at any time, even during the growth of a culture for the preparation of plasmid DNA. Thus, while organisms that generate totally different plasmid fingerprints certainly have totally different plasmid complements, and may therefore be epidemiologically distinct, isolates that generate fingerprints which differ by only one or two DNA fragments should be considered as having a possible close relationship unless supplementary data obtained by other typing methods are available.

2.3 ANALYSIS OF CHROMOSOMAL DNA

Although analysis of plasmid profiles and fingerprints can both be extremely useful techniques for microbial identification and typing, not all isolates carry plasmids (and are therefore not amenable to this form of analysis), while the plasmids themselves may either undergo molecular re-arrangements, or may be acquired or lost in a particular environment. Fortunately from a typing point of view, molecular re-arrangements seem to be relatively rare, but the possibility of a change must always be considered. Such problems can be overcome by examining the chromosomal DNA. However, the conventional agarose gel electrophoresis methods described earlier in this chapter are incapable of separating DNA molecules greater than 50–100 kb in size, while microbial chromosomes are typically several megabases (Mb) in size.

2.3.1 GENERATION OF CHROMOSOMAL FINGERPRINTS

One way to overcome the problem of size is to digest the chromosomal DNA with a restriction endonuclease to yield DNA fragments that are capable of resolution by standard electrophoresis methods. This method is sometimes referred to as BRENDA (bacterial restriction endonuclease digest analysis). BRENDA enables minor genomic differences between microorganisms to be detected directly. However, in practice, it is normal for either too many fragments to be produced, resulting in preparations where it is difficult to distinguish individual bands (and hence characteristic fingerprints), or else too few fragments are produced, which are difficult to electrophorese by conventional techniques because of their size. The utility of the procedure therefore depends crucially on the selection of an appropriate restriction endonuclease(s) and on the method of determining similarities between the fingerprints that are generated. Table 2.7 lists some of the organisms with which BRENDA has proved useful for epidemiological and typing purposes.

Since the genomes of microorganisms vary widely in their GC base compositions (25–75 mol % G+C), the fragments produced by cutting at specific restriction endonuclease digestion sites can also be expected to vary considerably. It is possible to calculate the expected frequency (a) of restriction sites from the following formula (Nei and Li, 1979):

$$a = (g/2)^{r1} \times [1-(g/2)]^{r2}$$

Table 2.7 Examples of microorganisms investigated by analysis of chromosomal fingerprints

Organism	Reference
Acholeplasma spp.	Razin *et al.* (1983)
Bacteroides ureolyticus	Akhtar and Eley (1992)
Campylobacter jejuni	Bradbury *et al.* (1984)
Candida albicans	Vasquez *et al.* (1991)
Chlamydia psittaci	McClenaghan *et al.* (1984)
Clavibacter michiganensis	Finnen *et al.* (1990)
Corynebacterium diphtheriae	Groman *et al.* (1983)
Enterobacter cloacae	Haertl and Bandlow (1993)
Enterococcus spp.	Hall *et al.* (1992)
Escherichia coli	Hampson *et al.* (1986)
Frankia spp.	Dobritsa (1985)
Giardia spp.	Nash *et al.* (1985)
Haemophilus influenzae	Bruce and Jordens (1991)
Human Polyoma JC Virus	Yogo *et al.* (1991)
Micromonospora spp.	Crameri *et al.* (1983)
Mycoplasma spp.	Chan and Ross (1984)
Neisseria meningitidis	Bjorvatn *et al.* (1984)
Nocardia spp.	Crameri *et al.* (1983)
Pasteurella multocida	Wilson *et al.* (1992)
Pseudomonas aeruginosa	Loutit and Tompkins (1991)
Rhizobium spp.	Mielenz *et al.* (1979)
Rickettsia spp.	Regnery *et al.* (1983)
Salmonella typhi	Franco *et al.* (1992)
Spiroplasma citri	Bové *et al.* (1982)
Staphylococci, coagulase-negative	Bialkowska-Hobrzanska *et al.* (1990)
Streptococci, group B	Denning *et al.* (1989)
Streptococcus pyogenes	Bingen *et al.* (1992)
Streptococcus uberis	Hill and Leigh (1989)
Streptomyces coelicolor	Crameri *et al.* (1983)
Ureaplasma spp.	Razin *et al.* (1983)
Vibrio cholerae	Kaper *et al.* (1982)

where g is the fractional G+C content of the genome DNA, $r1$ is the number of G+C base pairs, and $r2$ is the number of A+T base pairs in the recognition site of the particular restriction enzyme. The expected number of restriction sites in the genome is then calculated from the frequency (a) times the genome size.

Although the large genome sizes of microorganisms make it impossible to predict accurately how often a particular enzyme will cut a particular genome, such calculations do provide some information about probable cutting frequencies. Thus, the probability of a G+C-rich site is only about 1 in 80 000 in a G+C-poor DNA, compared with about 1 in 500 in a G+C-rich DNA.

The occurrence of particular sets of nucleotides in genomes and enzyme recognition sequences has been shown by McClelland *et al.*

(1987) to be a useful indicator of cutting frequencies. Similarly, Forbes *et al.* (1991) described a procedure for selecting a suitable restriction enzyme by calculating the frequencies of enzyme recognition sites from published DNA sequences for the species of interest. Subsequent analysis of fingerprints is achieved by dividing the fragments into size 'sections' by means of DNA size standards, scoring the number of bands in each section, and comparing these scores (numerical profiles) to establish similarities. In this way, a single conventional electrophoretic gel can yield data that can be compared with data from other gels. Alternative approaches have been described in which scoring and classification of fingerprints is achieved by scanning densitometry and computer analysis (Sorensen *et al.*, 1985; Owen *et al.*, 1990; Stahl *et al.*, 1990), but such approaches do not allow inter-gel comparisons to be made readily.

For most typing purposes in diagnostic laboratories, it is normal to make simple visual comparisons of band patterns to see if the overall fingerprints are similar or different. However, the chromosome of a microorganism will be reduced in molecular size by as much as 200-fold simply as a result of enzymic degradation and shearing forces during isolation and purification. This results from the large size of the original DNA molecule comprising the chromosome (typically between 800 and 8000 kb for most microorganisms), and is unavoidable when using conventional nucleic acid isolation and purification techniques. Digests of such DNA will, therefore, contain a high proportion of random fragments as well as those cut specifically by the particular restriction enzyme being used. This may influence the extent to which a distinct pattern of fragment bands can be visualized, and often a 'smear' will be observed with no obvious band pattern at all.

The presence of extrachromosomal plasmid or bacteriophage DNA can also complicate the interpretation of restriction endonuclease fingerprints. It is, therefore, advisable first to analyse undigested DNA by electrophoresis on agarose gels to test for the presence of any bands additional to that of the chromosomal DNA.

2.3.2 PULSED-FIELD GEL ELECTROPHORESIS (PFGE)

(a) Basic principles

A solution to the problems of analysis caused by complex chromosomal fingerprint patterns is to use a 'rare-cutting' restriction endonuclease to generate only a limited number of large DNA fragments. As stated in the previous section, such large fragments cannot be separated readily by conventional agarose gel electrophoresis. The only technique available currently that is capable of separating DNA molecules in the 50 kb–12 Mb range physically is pulsed-field gel electrophoresis (PFGE), first described by Schwartz and Cantor (1984). Although a large number of variations have been described subsequently (see below), the theory of all pulsed-field methods is similar. Their whole purpose is to provide improved resolution of large DNA molecules compared with conventional agarose gel electrophoresis, so the gels and buffers used are the same in terms of their composition as for conventional methods, and the only real difference is in the way that the electrical field is applied.

Conventional gel electrophoresis methods are based on the ability of a gel matrix to sieve molecules, on the basis of size, under the influence of a unidirectional electrical field. In contrast, PFGE utilizes successive alternating electric fields which force the migrating DNA molecules to change continuously their direction of migration. Separation is based, probably, on the fact that larger DNA molecules change direction more slowly than the smaller molecules, resulting in separation by retardation of DNA molecules by size rather than sieving. However, the precise mechanism of separation of DNA molecules on agarose gels is by no means understood precisely, although it has been the subject of several intensive investigations (Slater and Noolandi, 1989; Stellwagen and Stellwagen, 1989; Viovy, 1989).

Following the first description of PFGE (Schwartz and Cantor, 1984), it was thought that the mobility of different DNA molecules was a linear function of their molecular size, and that resolution of a particular size range could be achieved by adjusting the 'pulse' time of the electric field (Smith and Condemine, 1990). However, other physical factors have also been shown to be important (e.g. Birren, Hood and Lai, 1989), and it is known that changes in parameters such as temperature, voltage, agarose concentration and ionic strength will act inter-dependently, but non-uniformly, on the mobilities of different sizes of DNA. Thus, changes in one parameter might affect resolution adversely, but resolution can be restored by making adjustments in another parameter. Lack of a full understanding of the effects of the individual parameters means that the choice is often empirical, needing several trial experiments for a particular application.

To achieve separation of a broad range of DNA sizes, such as is often required for typing purposes, it is a common practice to increase the 'pulse' time gradually over the course of the experiment. This gradual increase (termed **ramping**) has the effect of ensuring that each size range of DNA molecules is subjected to optimal separation conditions for at least part of the time. The agarose concentration used most widely is 1%, with the size and shape of gels being determined by the precise method and design of apparatus used (see below), but most pulsed-field methods use horizontal agarose gels in a submerged ('submarine') mode. Buffers are circulated continuously and cooled, normally to a

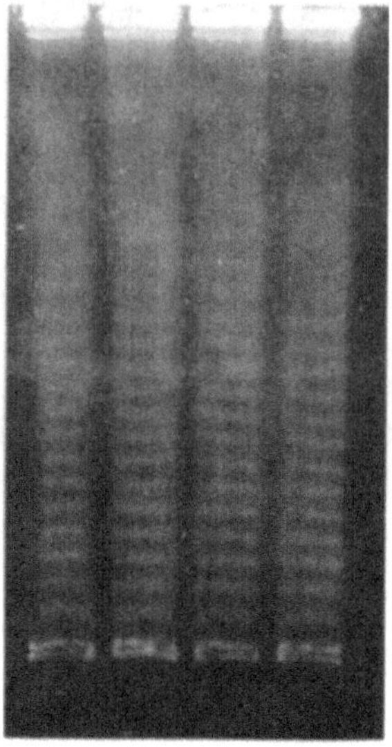

Figure 2.4 Pulsed-field gel electrophoresis of monomeric bacteriophage lambda concatamers. Each DNA band represents an increase in size of c. 48.5 kb.

constant temperature between 10° and 15°C. Electrical conditions and separation times depend on the precise method being used and the size range of the DNA molecules being separated, but for microbial identification and typing purposes would probably be within the range 2–10 V/cm for 6–30 h. It is important to note that attempts to speed-up separations by increasing voltages or reducing the agarose concentration often lead to reduced resolution. Following completion of electrophoresis, gels are stained with ethidium bromide (0.5 mg/l for 60 min), destained with distilled water as necessary, and the DNA visualized in the same manner as for conventional agarose gel electrophoresis.

Effective use of PFGE requires accurate and reliable size standards for estimating the sizes of the DNA fragments being studied. Figure 2.4 shows examples of concatamers of the monomeric bacteriophage lambda genome (48.5 kb) separated by PFGE over the size range from c. 49 to 800 kb. Each successive band differs in size by about 49 kb. Such markers are ideal for calculating the sizes of DNA fragments generated from the genome of prokaryotic organisms. Larger fragments generated from more complex microorganisms are often sized by comparison with the 16 chromosomes of *Saccharomyces cerevisiae*, spanning the size range from 225 to 1900 kb. Ready-prepared size markers for PFGE are available from several different commercial suppliers.

Table 2.8 Different types of PFGE

Abbreviation	Type of PFGE
UPFGE	Unidirectional pulsed-field gel electrophoresis (Sutherland *et al.*, 1987b; Chen and Sutherland, 1989)
OFAGE	Orthogonal field alternation gel electrophoresis (Carle and Olson, 1984)
FIGE	Field inversion gel electrophoresis (Carle *et al.*, 1986)
TAFE	Transverse alternating field electrophoresis (Gardiner *et al.*, 1986)
CHEF	Contour-clamped homogeneous electric fields (Chu *et al.*, 1986; Chu, 1989)
PACE	Programmable, autonomously-controlled electrode system (Clark, 1988; Birren *et al.*, 1989)
RGE	Rotating gel electrophoresis (Serwer, 1987; Southern *et al.*, 1987)

(b) Types of PFGE

Since the original description of PFGE, a number of different variations have been described (Table 2.8) which attempt to improve resolution and avoid distortions affecting the comparison of different DNA tracks. These techniques can be divided into two broad categories which either have a constant field orientation or which include a change in field orientation. Space does not permit the comparative advantages and disadvantages of the various methods to be discussed in this book, but the interested reader will find full details in a review by Andrews (1991). PFGE is an extremely active area of research, with further improvements and innovations being introduced continuously. However, at the present time, the most popular pulsed-field systems are probably those based on the contour-clamped homogeneous electric fields (CHEF) technique. Such systems are available commercially and rely on an hexagonal electrode array to produce a highly uniform electric field. The overall effect is to produce straight lanes and good resolution of DNA fragments up to and exceeding 5 Mb in size, although for the largest molecules it is necessary to use long run times (130 h) with very low field strengths (1.5 V/cm) and long pulse times (60 min). For 'smaller' DNA fragments of less than 2 Mb, 8 V/cm can be used with an 80 s pulse and 24 h run times.

(c) Preparation of DNA for PFGE

Large molecules of DNA are difficult to isolate by the rapid procedures used for plasmid DNA, since chromosome-sized DNA is extremely prone to random shearing during the normal preparative processes. Accordingly, methods have been developed which avoid liquid shearing forces by extracting chromosome-sized DNA molecules from intact cells embedded in low-melting-temperature (LMT) agarose plugs (Schwartz and Cantor, 1984; Smith, Klco and Cantor, 1988). In summary, the plugs are prepared by mixing a broth culture or suspension of the organisms being studied with LMT agarose (held briefly at 37°C) and pipetting the mixture into a perspex mould. When set, the plugs containing the intact cells are first treated with a combination of enzymes and detergents to lyse the cells *in situ* and remove cell walls, membranes, RNA and proteins. Cellular nucleases are inactivated by immersing the plugs in a solution containing a high concentration of EDTA, with proteinase K and *N*-lauroylsarcosine to remove all other cellular components. Portions of DNA-containing agarose plugs can then be loaded directly into the wells of 0.5–1.2% agarose gels, where they are sealed in place with molten LMT agarose before PFGE. An alternative approach involves remelting the plugs by heating to 65°C for 5 min, and then loading the DNA into the wells as a molten mixture.

Precise protocols for preparing DNA for PFGE vary according to whether or not the organism being studied has a cell wall. The detailed review of Smith *et al.* (1988) contains protocols for the preparation of chromosome-sized DNA from bacterial, fungal, protozoal and mammalian cells. In some cases (e.g. filamentous fungi), pre-treatment of cells with cell-wall degrading enzymes to generate spheroplasts or protoplasts is necessary before embedding them in LMT agarose. In any event, it is necessary to adjust empirically the number of cells used in order to yield between 0.5 and 20 μg of DNA per plug (Smith *et al.*, 1986, 1988). This is important since higher concentrations of DNA can result in loss of PFGE resolution. However, once agarose plugs containing appropriate amounts of DNA have been prepared, they can be stored at 4°C in solutions containing 0.5 M EDTA for at least a year.

For typing purposes (see below) it is necessary to digest chromosome-sized DNA molecules with 'rare-cutting' restriction endonucleases (see Table 2.9) in order to generate PFGE fingerprints. Agarose plugs containing free DNA must first be treated with phenylmethylsulphonyl fluoride (PMSF) in order to inactivate any remaining proteinase K (Smith *et al.*, 1988). In turn, the PMSF must then be removed by successive washings with detergents and EDTA. Small pieces of agarose plug can then be incubated overnight in microfuge tubes containing digestion buffer and an appropriate restriction endonuclease. The endonuclease

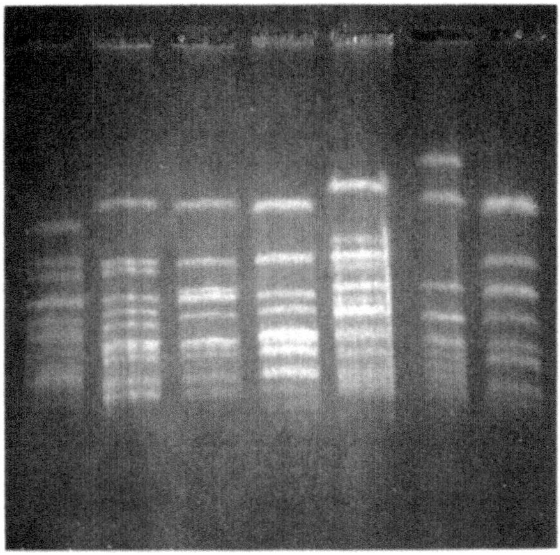

Figure 2.5 Pulsed-field gel electrophoresis of *Sma*I-digested chromosomal DNA from seven strains of *Haemophilus influenzae*.

diffuses into the agarose and digestion occurs *in situ*. Detailed protocols for the treatment of DNA in agarose plugs with restriction endonucleases can be found in Smith *et al*. (1986, 1988). Figure 2.5 shows examples of chromosomal DNA profiles obtained following digestion with a 'rare-cutting' restriction endonuclease and separation by PFGE.

Table 2.9 'Rare-cutting' restriction endonucleases suitable for use in conjunction with pulsed-field gel electrophoresis

Enzyme	Recognition sequence*
*Aat*II	GACGT/C
*Apa*I	GGGCC/C
*Cla*I	AT/CGAT
*Mlu*I	A/CGCGT
*Nar*I	GG/CGCC
*Nhe*I	G/CTAGC
*Not*I	GC/GGCCGC
*Nru*I	TCG/CGA
*Pvu*I	CGAT/CG
*Sac*II	CCGC/GG
*Sal*I	G/TCGAC
*Sfi*I	GGCC $(N)_4$/NGGCC
*Sma*I	CCC/GGG
*Xho*I	C/TCGAG

* N indicates a random nucleotide base; / indicates cutting site in relation to the recognition sequence.

(d) Applications of PFGE analysis for typing purposes

Restriction endonuclease analysis of the genome of microorganisms has long had the potential to be a general typing method for microorganisms, but has been hampered by the fact that digestions with frequently-cutting restriction enzymes yield a complex pattern of poorly resolved fragments. PFGE analysis of genomes digested with 'rare-cutting' enzymes overcomes this problem, and the technique has already been used for epidemiological and typing investigations with a wide range of microorganisms (Table 2.10). The pattern of restriction fragments is characteristic for each strain, and provides an estimate of the degree of genomic relationship between strains. Closely related strains that differ by only a few bands can be identified readily by side-to-side visual comparison of the fingerprint patterns on the same gel. The discriminatory value and information generated by the technique can be increased

Table 2.10 Examples of microorganisms investigated by PFGE of chromosomal DNA fragments

Organism	Reference
Acinetobacter baumannii	Gouby *et al.* (1992)
Brucella spp.	Allardet-Servent *et al.* (1988)
Campylobacter hyointestinalis	Salama *et al.* (1992)
Campylobacter jejuni	Yan *et al.* (1991)
Candida albicans	Vasquez *et al.* (1991)
Candida parapsilosis	Carruba *et al.* (1991)
Coxiella burnettii	Heinzen *et al.* (1990)
Enterobacter cloacae	Haertl and Bandlow (1993)
Enterococcus faecium	Miranda *et al.* (1991)
Escherichia coli	Bohm and Karch (1992)
Lactococcus spp.	Le Bourgeois *et al.* (1989)
Legionella pneumophila	Schoonmaker *et al.* (1992)
Leptospira spp.	Herrmann *et al.* (1992)
Listeria monocytogenes	Brosch *et al.* (1991)
Mycobacterium avium	Coffin *et al.* (1992)
Mycobacterium fortuitum	Hector *et al.* (1992)
Mycobacterium tuberculosis	Zhang *et al.* (1992)
Neisseria meningitidis	Bygraves and Maiden (1992)
Pseudomonas aeruginosa	Boukadida *et al.* (1991)
Pseudomonas cepacia	Anderson *et al.* (1991)
Other *Pseudomonas* spp.	Grothues and Tümmler (1991)
Schizosaccharomyces pombe	Smith *et al.* (1987)
Shigella spp.	Soldati and Piffaretti (1991)
Staphylococcus aureus	Prevost *et al.* (1992)
Staphylococci, coagulase-negative	Lina *et al.* (1992)
Streptococci, group A	Single and Martin (1992)
Torulopsis glabrata	Khattak *et al.* (1992)
Ureaplasma spp.	Robertson *et al.* (1990)

further by the use of more than one rare-cutting enzyme, either individually or collectively.

For some microorganisms with multiple chromosomes, e.g. *Candida albicans*, it may not even be necessary to digest the total genomic DNA with restriction enzymes. Monod *et al.* (1990) have proposed a karyotyping system for *C. albicans* in which electrophoretic karyotypes were assigned to strains on the basis of the number of bands that could be resolved by PFGE in each of four sets of chromosomes. Different chromosome preparations of the same strain gave identical band patterns under the same conditions of electrophoresis, and the method (based on the CHEF procedure) appeared to be a simple and reproducible method of differentiation at and below the species level.

A few caveats are, however, necessary. Firstly, analysis by PFGE requires specialized apparatus and is fairly skill-intensive, while the initial extraction of high molecular mass chromosomal DNA may be a difficult and time-consuming procedure with certain microorganisms. Secondly, PFGE is a relatively insensitive procedure for detecting small differences between strains. Thus, identical PFGE patterns obtained with a particular enzyme do not necessarily mean that two strains are identical. Differences are conclusive, but similarities are not. Caution should, however, be exercised when single band differences are observed. Such changes may result simply from genetic switching mechanisms rather than real differences between strains. As with chromosomal fingerprints obtained by conventional electrophoresis (section 2.3.1), the possibility of additional bands being caused by the presence of extrachromosomal plasmid or bacteriophage DNA should also be considered. Finally, it is important to note that PFGE tends to reflect the gross clonal structure of a collection of strains, rather than the 'fine architecture' which may be of more interest to epidemiologists. This also has the consequence that different 'rare-cutting' enzymes may appear to generate different typing results.

Despite these caveats, it seems that genomic fingerprinting by PFGE is a reliable technique that generates reproducible results. Grothues and Tümmler (1991) found that repetitive processing and macrorestriction analysis of more than 50 isolates of *Pseudomonas* spp. resulted in reproducible fragment patterns being generated for each strain analysed. Provided that accurate determination of fragment sizes can be achieved, fingerprinting results obtained in different laboratories can be compared directly. The gradual introduction of more rapid techniques for the isolation of intact chromosomal DNA from an increasing variety of microorganisms (e.g. Flanagan, Ventra and Weiss, 1989; Sobral and Atherley, 1989; Goering and Winters, 1992) means that classification and typing by analysis of PFGE fingerprints is rapidly becoming a suitable general technique that deserves consideration for investigating the epidemiology of most microorganisms.

2.4 SUMMARY

Conventional plasmid analysis is a technique that can be performed readily with simple equipment in most diagnostic microbiology laboratories. When combined with restriction endonuclease digestion of plasmid DNA to generate plasmid fingerprints, it can, with judicious interpretation, provide useful information on the degree of similarity of multiple isolates within a few hours. There is no need to keep sets of antisera or standard type organisms. Plasmids can, of course, be gained or lost from a particular organism, and it should always be kept in mind that plasmid profiles and fingerprints may lack stability with the passage of time. However, such events seem to be rare and, providing suitable plasmids are carried by the organisms being studied, plasmid analysis and/or fingerprinting enables local cross-infection problems to be identified almost as soon as they develop. It is worth noting that PFGE may be useful for the accurate separation and characterization of large plasmids which cannot be accurately sized or separated by conventional electrophoresis techniques.

One approach to the problem of typing organisms that lack suitable plasmids is to examine their chromosomal DNA. Restriction endonuclease analysis of genomic DNA isolated from microorganisms is now used widely, and the same basic technique can be applied to all microorganisms which are capable of being cultured. By means of a suitable 'rare-cutting' restriction endonuclease, it is possible to produce a small number of relatively large DNA fragments from the chromosome, and to analyse these by PFGE. Although this method can generate excellent results for typing purposes, the apparatus required is fairly expensive, while the techniques involved are more time-consuming and technically demanding. Hence this method is probably not, at present, suitable for large routine diagnostic laboratories with a high through-put of samples. It is also important to note that the use of a 'rare-cutting' restriction endonuclease results in a decrease in the number of restriction sites examined and, consequently, may reduce the ability to demonstrate differences between strains.

The alternative approach of conventional agarose gel electrophoresis demands that, in order to be separated, the chromosomal DNA is cut into a large number of relatively small fragments. Although this provides the basis of a good typing method, in practice it is difficult to distinguish individual bands in chromosomal DNA fingerprints because of their number and close proximity. This means that small differences may be missed easily. It is, however, possible to combine restriction analysis of chromosomal DNA with nucleic acid hybridization techniques to form the basis of a generalized microbial typing method. This method is the subject of the next chapter.

REFERENCES

Aaij, C. and Borst, P. (1972) The gel electrophoresis of DNA. *Biochimica et Biophysica Acta*, **269**, 192–200.

Abeck, D., Johnson, A.P., Dangor, Y. and Ballard, R.C. (1988) Antibiotic susceptibilities and plasmid profiles of *Haemophilus ducreyi* isolates from Southern Africa. *Journal of Antimicrobial Chemotherapy*, **22**, 437–44.

Agata, N., Mori, M. and Imai, M. (1990) Evaluation of plasmid profile as an epidemiologic tool for food poisoning caused by *Bacillus cereus*. *Journal of the Food Hygenic Society of Japan*, **31**, 378–81.

Akhtar, N. and Eley, A. (1992) Restriction endonuclease analysis and ribotyping differentiate genital and nongenital strains of *Bacteroides ureolyticus*. *Journal of Clinical Microbiology*, **30**, 2408–14.

Allardet-Servent, A., Bourg, G., Ramuz, M. *et al.* (1988) DNA polymorphism in strains of the genus *Brucella*. *Journal of Bacteriology*, **170**, 4603–7.

Anderson, D.J., Kuhns, J.S., Vasil, M.L. *et al.* (1991) DNA fingerprinting by pulsed field gel electrophoresis and ribotyping to distinguish *Pseudomonas cepacia* isolates from a nosocomial outbreak. *Journal of Clinical Microbiology*, **29**, 648–9.

Andrews, A.T. (1991) Electrophoresis of nucleic acids, in *Essential Molecular Biology: a Practical Approach*, vol. II (ed T.A. Brown), IRL Press, Oxford, pp. 89–126.

Barbour, A.G. and Garon, C.F. (1987) Linear plasmids of the bacterium *Borrelia burgdorferi* have covalently closed ends. *Science*, **237**, 409–11.

Bennett, P.M., Heritage, J. and Hawkey, P.M. (1986) An ultrarapid method for the study of antibiotic resistance plasmids. *Journal of Antimicrobial Chemotherapy*, **18**, 421–4.

Beul, H.A., von Eichel-Streiber, C.H., Schreiner, M. *et al.* (1985) Characterisation of cryptic plasmids in clinical isolates of *Bacteroides fragilis*. *Journal of Medical Microbiology*, **20**, 39–48.

Bialkowska-Hobrzanska, H., Jaskot, D. and Hammerberg, O. (1990) Evaluation of restriction endonuclease fingerprinting of chromosomal DNA and plasmid profile analysis for characterization of multiresistant coagulase-negative staphylococci in bacteremic neonates. *Journal of Clinical Microbiology*, **28**, 269–75.

Bingen, E., Denamur, E., Lambert-Zechovsky, N. *et al.* (1992) Mother-to-infant vertical transmission and cross-colonization of *Streptococcus pyogenes* confirmed by DNA restriction fragment length polymorphism analysis. *Journal of Infectious Diseases*, **165**, 147–50.

Birnboim, H.C. and Doly, J. (1979) A rapid alkaline extraction procedure for screening recombinant plasmid DNA. *Nucleic Acids Research*, **7**, 1513–23.

Birren, B.W., Hood, L. and Lai, E. (1989) Pulsed field gel electrophoresis – studies of DNA migration made with the programmable, autonomously-controlled electrode electrophoresis system. *Electrophoresis*, **10**, 302–9.

Bjorvatn, B., Lund, V., Kristiansen, B. *et al.* (1984) Application of restriction endonuclease fingerprinting of chromosomal DNA of *Neisseria meningitidis*. *Journal of Clinical Microbiology*, **19**, 763–5.

Bohm, H. and Karch, H. (1992) DNA fingerprinting of *Escherichia coli* O157:H7 strains by pulsed-field gel electrophoresis. *Journal of Clinical Microbiology*, **30**, 2169–72.

Bottone, E.J., Janda, J.M., Chiesa, C. *et al.* (1985) Assessment of plasmid profile, exoenzyme activity, and virulence in recent human isolates of *Yersinia enterocolitica*. *Journal of Clinical Microbiology*, **22**, 449–51.

Boukadida, J., de Montalembert, M., Gaillard, J.-L. *et al.* (1991) Outbreak of gut colonization by *Pseudomonas aeruginosa* in immunocompromised children undergoing total digestive decontamination: analysis by pulsed-field electrophoresis. *Journal of Clinical Microbiology*, **29**, 2068–71.

Bouvet, P.J.M. (1991) Typing of *Acinetobacter*, in *The Biology of Acinetobacter: Taxonomy, Clinical Importance, Molecular Biology, Physiology, Industrial Relevance*, (eds K.J. Towner, E. Bergogne-Bérézin and C.A. Fewson), Plenum Press, New York, pp. 37–51.

Bové, J.M., Saillard, C., Junca, P. *et al.* (1982) Guanine plus cytosine content, hybridization percentages and *Eco*RI restriction enzyme profiles of spiroplasmal DNA. *Reviews of Infectious Diseases*, **4**, S129–36.

Boyce, J.R. and Morter, R.L. (1986) Plasmid profile analysis of bovine isolates of *Pasteurella haemolytica*. *American Journal of Veterinary Research*, **47**, 1204–6.

Bradbury, W.C., Pearson, A.D., Marko, M.A. *et al.* (1984) Investigation of a *Campylobacter jejuni* outbreak by serotyping and chromosomal restriction endonuclease analysis. *Journal of Clinical Microbiology*, **19**, 342–6.

Broach, J.R. (1981) The yeast plasmid 2 μ circle, in *The Molecular Biology of The Yeast Saccharomyces: Life Cycle and Inheritance* (eds J.N. Strathern, E.W. Jones and J.R. Broach), Cold Spring Harbor Laboratory, New York, pp. 445–70.

Brosch, R., Buchrieser, C. and Rocourt, J. (1991) Subtyping of *Listeria monocytogenes* serovar–4b by use of low-frequency-cleavage restriction endonucleases and pulsed field gel electrophoresis. *Research in Microbiology*, **142**, 667–75.

Bruce, K.D. and Jordens, J.Z. (1991) Characterization of noncapsulate *Haemophilus influenzae* by whole-cell polypeptide profiles, restriction endonuclease analysis, and rRNA gene restriction patterns. *Journal of Clinical Microbiology*, **29**, 291–6.

Bygraves, J.A. and Maiden, M.C.J. (1992) Analysis of the clonal relationships between strains of *Neisseria meningitidis* by pulsed field gel electrophoresis. *Journal of General Microbiology*, **138**, 523–31.

Carle, G.F. and Olson, M.V. (1984) Separation of chromosomal DNA molecules from yeast by orthogonal field alternation gel electrophoresis. *Nucleic Acids Research*, **12**, 5647–64.

Carle, G.F., Frank, M. and Olson, M.V. (1986) Electrophoretic separations of large DNA molecules by periodic inversion of the electric field. *Science*, **232**, 65–8.

Carruba, G., Pontieri, E., De Bernardis, F. *et al.* (1991) DNA fingerprinting and electrophoretic karyotype of environmental and clinical isolates of *Candida parapsilosis*. *Journal of Clinical Microbiology*, **29**, 916–22.

Chan, H.W. and Ross, R.F. (1984) Restriction endonuclease analyses of 2 porcine mycoplasma deoxyribonucleic acids – sequence-specific methylation in the *Mycoplasma hyopneumoniae* genome. *International Journal of Systematic Bacteriology*, **34**, 16–20.

Chen, C.Z. and Sutherland, J.C. (1989) Gel electrophoresis method for quantitation of gamma-ray induced single-strand and double-strand breaks in DNA irradiated in vitro. *Electrophoresis*, **10**, 318–26.

Chu, G. (1989) Pulsed-field electrophoresis in contour-clamped homogeneous electric fields for the resolution of DNA by size and topology. *Electrophoresis*, **10**, 290–5.

Chu, G., Vollrath, D. and Davis, R.W. (1986) Separation of large DNA molecules by contour-clamped homogeneous electric fields. *Science*, **234**, 1582–5.

Clark, S.M., Lai, E., Birren, B.W. and Hood, L. (1988) A novel instrument for separating large DNA molecules with pulsed homogeneous electric fields. *Science*, **241**, 1203–5.

Coffin, J.W., Condon, C., Compston, C.A. *et al.* (1992) Use of restriction fragment length polymorphisms resolved by pulsed-field gel electrophoresis for subspecies identification of mycobacteria in the *Mycobacterium avium* complex and for isolation of DNA probes. *Journal of Clinical Microbiology*, **30**, 1829–36.

Coia, J. E., Noorhussain, I. and Platt, D. J. (1988) Plasmid profiles and restriction enzyme fragmentation patterns of plasmids of methicillin-sensitive and methicillin-resistant isolates of *Staphylococcus aureus* from hospital and the community. *Journal of Medical Microbiology*, **27**, 271–6.

Crameri, R., Hintermann, G. and Hutter, R. (1983) Deoxyribonucleic acid restriction endonuclease fingerprint characterization of actinomycete strains. *International Journal of Systematic Bacteriology*, **33**, 652–5.

Davies, F.L., Underwood, H.M. and Gasson, M.J. (1981) The value of plasmid profiles for strain identification in lactic streptococci and the relationship between *Streptococcus lactis* 712, ML3 and C2. *Journal of Applied Bacteriology*, **51**, 325–37.

Degrandis, S.A. and Stevenson, R.M.W. (1982) Variations in plasmid profiles and growth characteristics of *Yersinia ruckeri* strains. *FEMS Microbiology Letters*, **15**, 199–202.

Denning, D.W., Baker, C.J., Troup, N. and Tompkins, L.S. (1989) Restriction endonuclease analysis of human and bovine group B streptococci for epidemiologic study. *Journal of Clinical Microbiology*, **27**, 1352–6.

Dobritsa, S.V. (1985) Restriction analysis of the *Frankia* ssp. genome. *FEMS Microbiology Letters*, **29**, 123–8.

Dodd, C.E.R., Chaffey, B.J. and Waites, W.M. (1988) Plasmid profiles as indicators of the source of contamination of *Staphylococcus aureus* endemic within poultry-processing plants. *Applied and Environmental Microbiology*, **54**, 1541–9.

Domenico, P., Marx, J.L., Schoch, P.E. and Cunha, B.A. (1992) Rapid plasmid DNA isolation from mucoid Gram-negative bacteria. *Journal of Clinical Microbiology*, **30**, 2859–63.

Duggleby, R.G., Kinns, H. and Rood, J.I. (1981) A computer program for determining the size of DNA restriction fragments. *Analytical Biochemistry*, **110**, 49–55.

Dunkle, L.M. and Sippel, C.J. (1984) Rapid microprocedure for extraction of plasmid DNA from *Staphylococcus aureus*. *Journal of Infectious Diseases*, **149**, 921–3.

Dworkin, B.M., Chodos, J.E., Fernandez, M.E. *et al.* (1991) Use of plasmid profiles in the investigation of a patient with *Helicobacter pylori* infection and peptic ulcer disease. *American Journal of Gastroenterology*, **86**, 354–6.

Elder, J.K. and Southern, E.M. (1983) Measurement of DNA length by gel electrophoresis. II. Comparison of methods for relating mobility to fragment length. *Analytical Biochemistry*, **128**, 227–31.

Ellis, W.A., Montgomery, J.M. and Thiermann, A.B. (1991) Restriction endonuclease analysis as a taxonomic tool in the study of pig isolates belonging to the Australis serogroup of *Leptospira interrogans*. *Journal of Clinical Microbiology*, **29**, 957–61.

Ferdows, M.S. and Barbour, A.G. (1989) Megabase-sized linear DNA in the bacterium *Borrelia burgdorferi*, the Lyme disease agent. *Proceedings of the National Academy of Sciences of the USA*, **86**, 5969–73.

Finnen, R.L., Dhanvantari, B.N. and Trevors, J.T. (1990) Plasmid profiles and restriction endonuclease analysis of genomic DNA of *Clavibacter michiganensis* ssp *michagenensis*. *Journal of Microbiological Methods*, **12**, 57–64.

Fistrovici, E. and Collins-Thompson, D.L. (1990) Use of plasmid profiles and restriction endonuclease digest in environmental studies of *Listeria* spp. from raw milk. *International Journal of Food Microbiology*, **10**, 43–50.

Flanagan, J.L., Ventra, L. and Weiss, A.S. (1989) Rapid method for preparation and cleavage of bacterial DNA for pulsed-field gel electrophoresis. *Nucleic Acids Research*, **17**, 814.

Forbes, K.E., Bruce, K.D., Jordens, J.Z. *et al.* (1991) Rapid methods in bacterial DNA fingerprinting. *Journal of General Microbiology*, **137**, 2051–8.

Franco, A., Gonzalez, C., Levine, O.S. *et al.* (1992) Further consideration of the clonal nature of *Salmonella typhi*: evaluation of molecular and clinical characteristics of strains from Indonesia and Peru. *Journal of Clinical Microbiology*, **30**, 2187–90.

Friedrich, K.-H. and Lütticken, R. (1984) A rapid screening method for streptococcal plasmids yielding plasmid DNA suitable for restriction enzyme analysis. *FEMS Microbiology Letters*, **25**, 183–6.

Gardiner, K., Laas, W. and Patterson, D. (1986) Fractionation of large mammalian DNA restriction fragments using vertical pulsed-field gradient gel electrophoresis. *Somatic Cell and Molecular Genetics*, **12**, 185–95.

Gebre-Yohannes, A. and Drasar, B.S. (1990) Plasmid profiles of antibiotic-resistant *Shigella dysenteriae* type–2, type–3, type–4, type–6 and type–7 isolated in Ethiopia during 1976–85. *Epidemiology and Infection*, **105**, 65–72.

Goering, R.V. and Winters, M.A. (1992) Rapid method for epidemiological evaluation of gram-positive cocci by field inversion gel electrophoresis. *Journal of Clinical Microbiology*, **30**, 577–80.

Goldman, D. and Merrill, C.R. (1982) Silver-staining of DNA in polyacrylamide gels – linearity and effect of fragment size. *Electrophoresis*, **3**, 24–6.

Gouby, A., Carlesnurit, M.J., Bouziges, N. *et al.* (1992) Use of pulsed-field gel electrophoresis for investigation of hospital outbreaks of *Acinetobacter baumannii*. *Journal of Clinical Microbiology*, **30**, 1588–91.

Grimont, F. and Grimont, P.A.D. (1991) DNA fingerprinting, in *Nucleic Acid Techniques in Bacterial Systematics* (eds E. Stackebrandt and M. Goodfellow), Wiley, Chichester, pp. 249–79.

Groman, N., Cianciotto, N., Bjorn, M. and Rabin, M. (1983) Detection and expression of DNA homologous to the *tox* gene in nontoxigenic isolates of *Corynebacterium diphtheriae*. *Infection and Immunity*, **42**, 48–56.

Grothues, D. and Tümmler, B. (1991) New approaches in genome analysis by pulsed-field gel electrophoresis – application to the analysis of *Pseudomonas* species. *Molecular Microbiology*, **5**, 2763–76.

Haertl, R. and Bandlow, G. (1993) Epidemiological fingerprinting of *Enterobacter cloacae* by small-fragment restriction endonuclease analysis and pulsed-field gel electrophoresis of genomic restriction fragments. *Journal of Clinical Microbiology*, **31**, 128–33.

Hall, L.M.C., Duke, B., Guiney, M. and Williams, R. (1992) Typing of *Enterococcus* species by DNA restriction fragment analysis. *Journal of Clinical Microbiology*, **30**, 915–9.

Hampson, D.J., Bettelheim, K.A., Winter, P.J. *et al.* (1986) A comparison of serotyping, BRENDA-typing and incompatibility grouping, and toxin testing of hemolytic *Escherichia coli* isolated from piglets before and after weaning. *New Zealand Veterinary Journal*, **34**, 101–3.

Hartmann, A. and Amarger, N. (1991) Genotypic diversity of an indigenous *Rhizobium meliloti* field population assessed by plasmid profiles, DNA fingerprinting and insertion sequence typing. *Canadian Journal of Microbiology*, **37**, 600–8.

Hartstein, A.I., Valvano, M.A., Morthland, V.H. *et al.* (1987) Antimicrobic susceptibility and plasmid profile analysis as identity tests for multiple blood isolates of coagulase-negative staphylococci. *Journal of Clinical Microbiology*, **25**, 589–93.

Hawkey, P.M., Bennett, P.M. and Hawkey, C.A. (1984) Cryptic plasmids in hospital isolates of *Providencia stuartii*. *Journal of Medical Microbiology*, **18**, 277–84.

Hawkey, P.M., Penner, J.L., Linton, A.H. *et al.* (1986) Speciation, serotyping, antimicrobial sensitivity and plasmid content of Proteeae from the environment of calf-rearing units in South West England. *Journal of Hygiene*, **97**, 405–17.

Hector, J.S.R., Pang, Y.J., Mazurek, G.H. *et al.* (1992) Large restriction fragment patterns of genomic *Mycobacterium fortuitum* DNA as strain-specific markers and their use in epidemiologic investigation of four nosocomial outbreaks. *Journal of Clinical Microbiology*, **30**, 1250–5.

Heinzen, R., Stiegler, G.L., Whiting, L.L. *et al.* (1990) Use of pulsed field gel electrophoresis to differentiate *Coxiella burnetii* strains. *Annals of the New York Academy of Sciences*, **590**, 504–13.

Herrmann, J.L., Bellenger, E., Perolat, P. *et al.* (1992) Pulsed-field gel electrophoresis of *Not*I digests of leptospiral DNA – a new rapid method of serovar identification. *Journal of Clinical Microbiology*, **30**, 1696–702.

Hill, A.W. and Leigh, J.A. (1989) DNA fingerprinting of *Streptococcus uberis*: a useful tool for epidemiology of bovine mastitis. *Epidemiology and Infection*, **103**, 165–71.

Hintermann, G., Fischer, H.M., Crameri, R. and Hutter, R. (1981) Simple procedure for distinguishing CCC, OC and L forms of plasmid DNA by agarose gel electrophoresis. *Plasmid*, **5**, 371–3.

Holmes, D.S. and Quigley, M. (1981) A rapid boiling method for the preparation of bacterial plasmids. *Analytical Biochemistry*, **114**, 193–7.

Hotta, K., Saito, N. and Okami, Y. (1980) Studies on new aminoglycoside antibiotics, istamycins, from an actinomycete isolated from a marine environment. 1. The use of plasmid profiles in screening antibiotic-producing streptomycetes. *Journal of Antibiotics*, **33**, 1502–9.

Huether, M.J., Fedorkacray, P.J., Pfannenstiel, M.A. and Anderson, G.A. (1987) Plasmid profiles and antibiotic susceptibility of *Hemophilus pleuropneumoniae* serotypes–1, serotypes–3, serotypes–5, and serotypes–7. *FEMS Microbiology Letters*, **48**, 179–82.

Ish-Horowicz, D. and Burke, J.F. (1981) Rapid and efficient cosmid cloning. *Nucleic Acids Research*, **9**, 2989–98.

John, J.F. and Twitty, J.A. (1986) Plasmids as epidemiologic markers in nosocomial Gram-negative bacilli: experience at a university and review of the literature. *Reviews of Infectious Diseases*, **8**, 693–704.

Kado, C.I. and Liu, S.-T. (1981) Rapid procedure for detection and isolation of large and small plasmids. *Journal of Bacteriology*, **145**, 1365–73.

Kaper, J.B., Bradford, H.B., Roberts, N.C. and Falkow, S. (1982) Molecular epidemiology of *Vibrio cholerae* in the United States gulf coast. *Journal of Clinical Microbiology*, **16**, 129–34.

Khattak, M.N., Burnie, J.P., Matthews, R.C. and Oppenheim, B.A. (1992) Clamped homogeneous electric field gel electrophoresis typing of *Torulopsis glabrata* isolates causing nosocomial infections. *Journal of Clinical Microbiology*, **30**, 2211–5.

Kieser, T. (1984a) Factors affecting the isolation of CCC DNA from *Streptomyces lividans* and *Escherichia coli*. *Plasmid*, **12**, 19–36.

Kieser, T. (1984b) DNAGEL – a computer program for determining DNA fragment sizes using a small computer equipped with a graphics tablet. *Nucleic Acids Research*, **12**, 679–88.

Kinashi, H., Shimaji, M. and Sakai, A. (1987) Giant linear plasmids in *Streptomyces* which code for antibiotic biosynthesis genes. *Nature*, **238**, 454–6.

Klaenhammer, T.R. (1984) A general method for plasmid isolation in lactobacilli. *Current Microbiology*, **10**, 23–8.

Klein, R.D., Selsing, E. and Wells, R.D. (1980) A rapid microscale technique for isolation of recombinant plasmid DNA suitable for restriction enzyme analysis. *Plasmid*, **3**, 88–91.

Lastovica, A.J. and Ambrosio, R.E. (1986) Plasmid profiles of clinical isolates of *Campylobacter* with no or weak catalase activity. *South African Medical Journal*, **70**, 337–8.

Le Bourgeois, P., Mata, M. and Ritzenthaler, P. (1989) Genome comparison of *Lactococcus* strains by pulsed-field electrophoresis. *FEMS Microbiology Letters*, **59**, 65–70.

Lina, B., Vandenesch, F., Etienne, J. *et al.* (1992) Comparison of coagulase-negative staphylococci by pulsed-field gel electrophoresis. *FEMS Microbiology Letters*, **92**, 133–8.

Loos, B., Bernstein, J.M., Dryja, D.M. *et al.* (1989) Determination of the epidemiology and transmission of nontypable *Haemophilus influenzae* in children with otitis media by comparison of total genomic DNA restriction fingerprints. *Infection and Immunity*, **57**, 2751–7.

Loutit, J.S. and Tompkins, L.S. (1991) Restriction enzyme and Southern hybridization analyses of *Pseudomonas aeruginosa* strains from patients with cystic fibrosis. *Journal of Clinical Microbiology*, **29**, 2897–900.

Maher, W.E., Plouffe, J.F. and Para, M.F. (1983) Plasmid profiles of clinical and environmental isolates of *Legionella pneumophila* serogroup–1. *Journal of Clinical Microbiology*, **18**, 1422–3.

McClelland, M. , Jones, R., Patel, Y. and Nelson, M. (1987) Restriction endonucleases for pulsed field mapping of bacterial genomes. *Nucleic Acids Research*, **15**, 5985–6005.

McClenaghan, M., Herring, A.J. and Aitken, I.D. (1984) Comparison of *Chlamydia psittaci* isolates by DNA restriction endonuclease analysis. *Infection and Immunity*, **45**, 384–7.

McDonald, T.J. and Pugh, G.W. (1986) Plasmid profiles of *Moraxella bovis* isolates. *American Journal of Veterinary Research*, **47**, 964–6.

Mee, B.J., Turbett, G. and Rich, G. (1988) The epidemiology of *Neisseria gonorrhoeae* isolates from Western Australia based on antibiotic resistance and plasmid profiles. *Journal of Antimicrobial Chemotherapy*, **21**, 117–24.

Meissner, P.S. and Falkinham, J.O. (1986) Plasmid DNA profiles as epidemiological markers for clinical and environmental isolates of *Mycobacterium avium*, *Mycobacterium intracellulare*, and *Mycobacterium scrofulaceum*. *Journal of Infectious Diseases*, **153**, 325–31.

Meyers, J.A., Sanchez, D., Elwell, L.P. and Falkow, S. (1976) Simple agarose gel electrophoretic method for the identification and characterization of plasmid deoxyribonucleic acid. *Journal of Bacteriology*, **127**, 1529–37.

Mielenz, J.R., Jackson, L.E., O'Gara, F. and Shanmugan, K.T. (1979) Fingerprinting bacterial chromosomal DNA with restriction endonuclease *Eco*RI – comparison of *Rhizobium* spp., and identification of mutants. *Canadian Journal of Microbiology*, **25**, 803–7.

Miranda, A.G., Singh, K.V. and Murray, B.E. (1991) DNA fingerprinting of *Enterococcus faecium* by pulsed-field gel electrophoresis may be a useful epidemiologic tool. *Journal of Clinical Microbiology*, **29**, 2752–7.

Monod, M., Porchet, S., Baudraz-Rosselet, F. and Frenk, E. (1990) The identification of pathogenic yeast strains by electrophoretic analysis of their chromosomes. *Journal of Medical Microbiology*, **32**, 123–9.

Mulligan, M.E., Peterson, L.R., Kwok, R.Y.Y. *et al*. (1988) Immunoblots and plasmid fingerprints compared with serotyping and polyacrylamide gel electrophoresis for typing *Clostridium difficile*. *Journal of Clinical Microbiology*, **26**, 41–6.

Nahaie, M.R., Goodfellow, M. and Harwood, C.R. (1984) A rapid screening procedure for staphylococcal plasmids. *Journal of Microbiological Methods*, **2**, 73–81.

Nash, T.E., McCutchan, T., Keister, D. *et al*. (1985) Restriction endonuclease analysis of DNA from 15 *Giardia* isolates obtained from humans and animals. *Journal of Infectious Diseases*, **152**, 64–73.

Nei, M. and Li, W.-H. (1979) Mathematical model for studying genetic variation in terms of restriction endonucleases. *Proceedings of the National Academy of Sciences of the USA*, **76**, 5269–73.

Noterdaeme, L., Bigawa, S., Willems, K.A. and Ollevier, F. (1991) Biochemical and physiological characteristics and plasmid profiles of *Aeromonas hydrophila* strains, isolated from fresh-water fish and from fresh-water. *Journal of Fish Diseases*, **14**, 313–21.

Old, R.W. and Primrose, S.B. (1985) Cutting and joining DNA molecules, in *Principles of Gene Manipulation*, Blackwell, Oxford, pp. 20–9.

Olsen, J.E. (1990) An improved method for rapid isolation of plasmid DNA from wild-type Gram-negative bacteria for plasmid restriction profile analysis. *Letters in Applied Microbiology*, **10**, 209–12.

Owen, R.J. and Beck, A. (1987) Evaluation of three procedures using a laser densitometer and microcomputer for estimating molecular sizes of restriction endonuclease digest fragments and application to *Campylobacter jejuni* chromosomal DNA. *Letters in Applied Microbiology*, **4**, 5–8.

Owen, R.J., Fraser, J., Costas, M. *et al.* (1990) Signature patterns of DNA restriction fragments of *Helicobacter pylori* before and after treatment. *Journal of Clinical Pathology*, **43**, 646–9.

Parisi, J.T. (1985) Coagulase-negative staphylococci and the epidemiological typing of *Staphylococcus epidermidis*. *Microbiological Reviews*, **49**, 126–39.

Platt, D.J. and Sommerville, J.S. (1981) *Serratia* species isolated from patients in a general hospital. *Journal of Hospital Infection*, **2**, 341–8.

Platt, D.J., Chesham, J.S., Brown, D.J. *et al.* (1986) Restriction enzyme fingerprinting of enterobacterial plasmids: a simple strategy with wide application. *Journal of Hygiene*, **97**, 205–10.

Plikaytis, B.D., Carlone, G.M., Edmonds, P. and Mayer, L.W. (1986) Robust estimation of standard curves for protein molecular weight and linear duplex DNA base pair number after gel electrophoresis. *Analytical Biochemistry*, **152**, 346–64.

Poh, C.L., Yap, E.H., Tay, L. and Bergan, T. (1988) Plasmid profiles compared with serotyping and pyocin typing for epidemiological surveillance of *Pseudomonas aeruginosa*. *Journal of Medical Microbiology*, **25**, 109–14.

Prevost, G., Jaulhac, B. and Piemont, Y. (1992) DNA fingerprinting by pulsed-field gel electrophoresis is more effective than ribotyping in distinguishing among methicillin-resistant *Staphylococcus aureus* isolates. *Journal of Clinical Microbiology*, **30**, 967–73.

Prunell, A., Kopecka, H., Strauss, F. and Bernardi, G. (1977) The mitochondrial genome of wild-type yeast cells. *Journal of Molecular Biology*, **110**, 17–52.

Razin, S., Tully, J.G., Rose, D.L. and Barile, M.F. (1983) DNA cleavage patterns as indicators of genotypic heterogeneity among strains of *Acholeplasma* and *Mycoplasma* species. *Journal of General Microbiology*, **129**, 1935–44.

Regnery, R.L., Tzianabos, T., Esposito, J.J. and McDade, J.E. (1983) Strain differentiation of epidemic typhus rickettsiae (*Rickettsia prowazeckii*) by DNA restriction endonuclease analysis. *Current Microbiology*, **8**, 355–8.

Riley, L.W., DiFerdinando, G.T., DeMelfi, T.M. and Cohen, M.L. (1983a) Evaluation of isolated cases of salmonellosis by plasmid profile analysis: introduction and transmission of a bacterial clone by precooked roast beef. *Journal of Infectious Diseases*, **148**, 12–7.

Riley, L.W., Remis, R.S., Helgerson, S.D. *et al.* (1983b) Hemorrhagic colitis associated with a rare *Escherichia coli* serotype. *New England Journal of Medicine*, **308**, 681–5.

Roberts, R.J. and Macelis, D. (1991) Restriction enzymes and their isoschizomers. *Nucleic Acids Research*, **19S**, 2077–109.

Robertson, J.A., Pyle, L.E., Stemke, G.W. and Finch, L.R. (1990) Human ureaplasmas show diverse genome sizes by pulsed-field electrophoresis. *Nucleic Acids Research*, **18**, 1451–5.

Rochelle, P.A., Fry, J.C., Day, M.J. and Bale, M.J. (1985) An accurate method for estimating sizes of small and large plasmids and DNA fragments by gel electrophoresis. *Journal of General Microbiology*, **132**, 53–9.

Salama, S.M., Tabor, H., Richter, M. and Taylor, D.E. (1992) Pulsed-field gel electrophoresis for epidemiologic studies of *Campylobacter hyointestinalis* isolates. *Journal of Clinical Microbiology*, **30**, 1982–4.

Schaberg, D.R. and Zervos, M. (1986) Plasmid analysis in the study of the epidemiology of nosocomial Gram-positive cocci. *Reviews of Infectious Diseases*, **8**, 705–12.

Schaberg, D.R., Tompkins, L.S. and Falkow, S. (1981) Use of agarose gel electrophoresis of plasmid deoxyribonucleic acid to fingerprint Gram-negative bacilli. *Journal of Clinical Microbiology*, **13**, 1105–8.

Schaffer, H.E. and Sederoff, R.R. (1981) Improved estimation of DNA fragment length from agarose gels. *Analytical Biochemistry*, **115**, 113–22.

Schoonmaker, D., Heimberger, T. and Birkhead, G. (1992) Comparison of ribotyping and restriction enzyme analysis using pulsed-field gel electrophoresis for distinguishing *Legionella pneumophila* isolates obtained during a nosocomial outbreak. *Journal of Clinical Microbiology*, **30**, 1491–8.

Schwan, T.G., Burgdorfer, W. and Garon, C.F. (1988) Changes in infectivity and plasmid profile of the Lyme disease spirochete, *Borrelia burgdorferi*, as a result of in vitro cultivation. *Infection and Immunity*, **56**, 1831–6.

Schwartz, D.C. and Cantor, C.R. (1984) Separation of yeast chromosome-sized DNAs by pulsed-field gradient gel electrophoresis. *Cell*, **37**, 67–75.

Scotland, S.M., Willshaw, G.A., Smith, H.R. *et al.* (1989) Adhesion to cells in culture and plasmid profiles of enteropathogenic *Escherichia coli* isolated from outbreaks and sporadic cases of infant diarrhea. *Journal of Infection*, **19**, 237–49.

Serwer, P. (1987) Gel electrophoresis with discontinuous rotation of the gel: an alternative to gel electrophoresis with changing direction of the electrical field. *Electrophoresis*, **8**, 301–4.

Single, L.A. and Martin, D.R. (1992) Clonal differences within M-types of the group A streptococcus revealed by pulsed field gel electrophoresis. *FEMS Microbiology Letters*, **91**, 85–90.

Slater, G.W. and Noolandi, J. (1989) Effect of non-parallel alternating fields on the mobility of DNA in the biased reptation model of gel electrophoresis. *Electrophoresis*, **10**, 413–28.

Smith, C.L. and Condemine, G. (1990) New approaches for physical mapping of small genomes. *Journal of Bacteriology*, **172**, 1167–72.

Smith, C.L., Warburton, P.E., Gall, A. and Cantor, C.R. (1986) Analysis of genome organization and rearrangements by pulsed-field gradient gel electrophoresis. *Genetic Engineering*, **8**, 45–70.

Smith, C.T., Matsumoto, T., Niwa, O. *et al.* (1987) An electrophoretic karyotype for *Schizosaccharomyces pombe* by pulsed-field electrophoresis. *Nucleic Acids Research*, **15**, 4481–9.

Smith, C.L., Klco, S.R. and Cantor, C.R. (1988) Pulsed-field gel electrophoresis and the technology of large DNA molecules, in *Genome Analysis: a Practical Approach* (ed K.E. Davis), IRL Press, Oxford, pp. 41–72.

Sobral, B.W.S. and Atherley, A.G. (1989) A rapid and cost-effective method for preparing genomic DNA from gram-negative bacteria in agarose plugs for pulsed-field gel electrophoresis. *BioTechniques*, **7**, 938.

Soldati, L. and Piffaretti, J.C. (1991) Molecular typing of *Shigella* strains using pulsed field gel electrophoresis and genome hybridisation with insertion sequences. *Research in Microbiology*, **142**, 489–98.

Sorensen, B., Falk, E.S., Wisloff-Nilsen, E. *et al.* (1985) Multivariate analysis of *Neisseria* DNA restriction endonuclease patterns. *Journal of General Microbiology*, **131**, 3099–100.

Southern, E.M. (1979) Measurement of DNA length by gel electrophoresis. *Analytical Biochemistry*, **100**, 319–23.

Southern, E.M., Anand, R., Brown, W.R.A. and Fletcher, D.S. (1987) A model for the separation of large DNA molecules by crossed field electrophoresis. *Nucleic Acids Research*, **15**, 5925–43.

Speyerer, P.D. and Boyle, J.A. (1987) The plasmid profile of *Edwardsiella ictaluri*. *Journal of Fish Diseases*, **10**, 461–9.

Spiliopoulou, I., Droukopoulou, A., Athanassiadou, A. and Dimitracopoulos, G. (1992) Plasmid profiles of *Acinetobacter* and *Enterobacter* species of hospital origin – restriction endonuclease analysis of plasmid DNA and transformation of *Escherichia coli* by R plasmids. *Journal of Chemotherapy*, **4**, 72–7.

Stahl, M., Molin, G., Persson, A. *et al.* (1990) Restriction endonuclease patterns and multivariate analysis as a classification tool for *Lactobacillus* spp. *International Journal of Systematic Bacteriology*, **40**, 189–93.

Stanisich, V.A. (1988) Identification and analysis of plasmids at the genetic level, in *Methods in Microbiology*, vol. 21 (eds J. Grinsted and P.M. Bennett), Academic Press, London, pp. 11–47.

Steinberg, J.P., Beckerdite, M.E. and Westenfelder, G.O. (1987) Plasmid profiles of *Clostridium difficile* isolates from patients with antibiotic-associated colitis in 2 community hospitals. *Journal of Infectious Diseases*, **156**, 1036–8.

Stellwagen, N.C. and Stellwagen, J. (1989) Orientation of DNA and the agarose gel matrix in pulsed electric fields. *Electrophoresis*, **10**, 332–44.

Sutherland, J.C., Lin, B., Monteleone, D.C. *et al.* (1987a) Electronic imaging system for direct and rapid quantitation of fluorescence from electrophoretic gels – application to ethidium bromide-stained DNA. *Analytical Biochemistry*, **163**, 446–57.

Sutherland, J.C., Monteleone, D.C., Mugavero, J.H. and Trunk, J. (1987b) Unidirectional pulsed-field electrophoresis of single-stranded and double-stranded DNA in agarose gels – analytical expressions relating mobility and molecular length and their application in the measurement of strand breaks. *Analytical Biochemistry*, **162**, 511–520.

Suwanto, A. and Kaplan, S. (1989) Physical and genetic mapping of the *Rhodobacter sphaeroides* 2.4.1 genome: presence of two unique circular chromosomes. *Journal of Bacteriology*, **171**, 5850–9.

Takahashi, S. and Nagano, Y. (1984) Rapid procedure for isolation of plasmid DNA and application to epidemiological analysis. *Journal of Clinical Microbiology*, **20**, 608–13.

Taylor, D.N., Wachsmuth, I.K., Shangkuan, Y.-H. *et al.* (1982) Salmonellosis associated with marijuana. *New England Journal of Medicine*, **306**, 1249–53.

Threlfall, E.J., Hall, M.L.M. and Rowe, B. (1986) *Salmonella gold-coast* from outbreaks of food-poisoning in the British Isles can be differentiated by plasmid profiles. *Journal of Hygiene*, **97**, 115–22.

Valenti, P., Polidoro, M., Buonfiglio, V. *et al.* (1990) Plasmid DNA profiles in *Thiobacillus ferooxidans*. *Journal of General and Applied Microbiology*, **36**, 351–5.

Vasquez, J.A., Beckley, A., Sobel, J.D. and Zervos, M.J. (1991) Comparison of restriction enzyme analysis and pulsed-field gradient gel electrophoresis as typing systems for *Candida albicans*. *Journal of Clinical Microbiology*, **29**, 962–7.

Viovy, J.L. (1989) Reptation breathing theory of pulsed electrophoresis – dynamic regimes, antiresonance and symmetry breakdown effects. *Electrophoresis*, **10**, 429–41.

Vollrath, D. and Davis, R.W. (1987) Resolution of DNA molecules greater than 5 megabases by contour-clamped homogeneous electric fields. *Nucleic Acids Research*, **15**, 7865–75.

von Husby, K.O. and Nes, I.F. (1986) Changes in the plasmid profile of *Lactobacillus plantarum* obtained from commercial meat starter cultures. *Journal of Applied Bacteriology*, **60**, 413–7.

Wachsmuth, I.K., Kiehlbauch, J.A., Bopp, C.A. *et al.* (1991) The use of plasmid profiles and nucleic acid probes in epidemiologic investigations of food-borne diarrheal diseases. *International Journal of Food Microbiology*, **12**, 77–90.

Wang, C.-C., Chu, M.-L., Ho, L.-J., and Hwang, R.-C. (1991) Analysis of plasmid pattern in paediatric intensive care unit outbreaks of nosocomial infection due to *Enterobacter cloacae*. *Journal of Hospital Infection*, **19**, 33–40.

Wells, J.G., Davis, B.R., Wachsmuth, I.K. *et al.* (1983) Laboratory investigation of hemorrhagic colitis outbreaks associated with a rare *Escherichia coli* serotype. *Journal of Clinical Microbiology*, **18**, 512–20.

Wiik, R., Andersen, K., Daae, F.L. and Hoff, K.A. (1989) Virulence studies based on plasmid profiles of the fish pathogen *Vibrio salmonicida*. *Applied and Environmental Microbiology*, **55**, 819–25.

Williams, W.W., Mariano, J., Spurrier, M. *et al.* (1984) Nosocomial meningitis due to *Citrobacter diversus* in neonates: new aspects of the epidemiology. *Journal of Infectious Diseases*, **150**, 229–35.

Wilson, C.R., Totten, P.A. and Baldwin, J.N. (1978) Rapid procedure for the detection of plasmids in *Staphylococcus epidermidis*. *Applied and Environmental Microbiology*, **36**, 368–74.

Wilson, M.A., Rimler, R.B. and Hoffman, L.J. (1992) Comparison of DNA fingerprints and somatic serotypes of serogroup B and E *Pasteurella multocida* isolates. *Journal of Clinical Microbiology*, **30**, 1518–24.

Yablonsky, M.D., Goodman, A.E., Stevnsborg, N. *et al.* (1988) *Zymomonas mobilis* CP4 – a clarification of strains via plasmid profiles. *Journal of Biotechnology*, **9**, 71–9.

Yan, W., Chang, N. and Taylor, D.E. (1991) Pulsed-field gel electrophoresis of *Campylobacter jejuni* and *Campylobacter coli* genomic DNA and its epidemiologic application. *Journal of Infectious Diseases*, **163**, 1068–72.

Yogo, Y., Iida, T., Taguchi, F. *et al.* (1991) Typing of human polyomavirus JC virus on the basis of restriction fragment length polymorphisms. *Journal of Clinical Microbiology*, **29**, 2130–8.

Zhang, Y.S., Mazurek, G.H., Cave, M.D. *et al.* (1992) DNA polymorphisms in strains of *Mycobacterium tuberculosis* analyzed by pulsed-field gel electrophoresis – a tool for epidemiology. *Journal of Clinical Microbiology*, **30**, 1551–6.

Identification and typing by nucleic acid hybridization techniques

3

3.1 INTRODUCTION

Hybridization tests utilizing nucleic acid probes are now used extensively by many laboratories for the direct detection of specific microorganisms in a wide variety of samples. The most common approach involves the use as a probe of either a specific cloned DNA fragment or a synthetic oligonucleotide (section 3.2.1). Both types of probe can identify a target organism (sometimes by detecting a specific gene conferring virulence) with high specificity in a polymicrobial environment. Many kits are now available commercially for performing such identification tests easily and rapidly, including kits for the detection of organisms that cannot readily be cultured *in vitro*.

An alternative approach, involving analysis of variations (RFLPs; restriction fragment length polymorphisms) between related DNA molecules, has already been alluded to in Chapter 2 as a method for comparing chromosomal DNA molecules for identification and typing purposes. In this procedure the DNA fragments generated by restriction endonuclease digestion are separated by electrophoresis in agarose gels to yield a pattern of bands for subsequent analysis. While it is sometimes possible to use 'rare-cutting' restriction endonucleases in combination with pulsed-field gel electrophoresis to generate useful epidemiological information (Chapter 2), in many cases the usefulness of genomic fingerprint patterns is limited by their great complexity; indeed, some enzymes may cut the DNA so frequently that it may be impossible to visualize individual bands.

One method of overcoming this problem is to use nucleic acid hybridization techniques to detect small specific regions of particular genomes that are themselves diagnostic of particular species. Alternatively, hybridization techniques can be used, in combination with a 'universal' probe, to examine small variations in conserved or semiconserved regions, and thereby generate a much simpler pattern of fragments for analysis.

This chapter outlines the basic principles of nucleic acid hybridization techniques, followed by a more detailed description of the analysis of RFLPs by hybridization methods for microbial identification and typing purposes.

3.2 BASIC PRINCIPLES OF NUCLEIC ACID HYBRIDIZATION

Hybridization reactions take advantage of the ability of nucleic acid molecules to form double strands in which nucleotides on opposing strands are held together by hydrogen bonds. When the hydrogen bonds are broken, normally by treatment with alkali or heating, the opposing strands separate and the nucleic acid molecule is said to be **denatured**. If a labelled 'probe' molecule is introduced at this stage, subsequent neutralization or cooling will allow the probe to **anneal** (i.e. form hydrogen bonds) with any complementary sequence that was present in the original denatured 'target' molecule. The specificity (**stringency**) and speed of such a hybridization reaction can be controlled by manipulating the temperature, salt concentration, pH, probe concentration, and probe size.

The concept of stringency is very important to understanding the specificity of hybridization reactions. If any of the parameters referred to above are changed, then the apparent specificity of the probe will also change. The greater the stringency, the greater the specificity, or 'match', that is required for successful hybridization to occur.

A positive hybridization reaction, after washing to remove excess unbound probe, is recognized by means of a detectable label attached to the hybridized probe molecules. Thus a hybridization reaction consists of four main components: (i) the probe molecule; (ii) the probe label ('reporter' molecule); (iii) the 'target' molecule (contained in the sample); and (iv) the precise hybridization format and stringency conditions.

3.2.1 PROBE CONSTRUCTION AND CHOICE

Most microbiologists interested in identification and typing methods will not be involved directly in the construction and development of nucleic acid probes. Interested readers are referred to the detailed review by Stahl and Amann (1991), but a brief overview is given here.

The basic requirements for a nucleic acid probe are that it should hybridize with the target nucleic acid molecule, but fail to hybridize with any other nucleic acid molecules which may be present in the sample being examined. By definition, all organisms contain at least one unique sequence of nucleic acid that serves to distinguish them from all other organisms and can, therefore, function as a probe. However, in practice, the selection or design of a suitable probe is often an empirical process, with a common approach involving random cross-screening of recombinant 'libraries' of total genomic nucleic acid to select for probes of the desired specificity. The final sequence selected need not be a whole gene, and can even be selected from outside a sequence that encodes a protein. A more rational, or directed, approach involves selecting a particular characteristic property of the organism being studied, cloning the gene encoding this property (which is often a specific surface epitope, toxin or plasmid-encoded function), and using part of the cloned gene as a probe.

Once a specific gene or sequence of DNA has been selected and analysed, it is then possible to synthesize very small stretches of nucleotides from within the overall sequence for use as probes. These short probes, termed **oligonucleotides**, are normally only 14–40 bp in length, and can be synthesized in the laboratory with one of several automated instruments designed specifically for this purpose. Oligonucleotides can display exquisite sensitivity when used as probes, such that, under stringent conditions, they may detect a change in a single base pair of a nucleic acid sequence (Wallace *et al.*, 1979).

Not only are oligonucleotide probes stable and simple to prepare, but also, because of their small size, they hybridize to their target extremely rapidly, often with reaction times of <30 min (Bryan *et al.*, 1986). This is a considerable advantage compared with larger cloned probes that often require up to 16 h for complete hybridization to occur. The main disadvantage of oligonucleotides as probes is that their small size means that they are sometimes more difficult to label by certain methods, resulting in a possible loss of sensitivity.

Perhaps the most powerful approach to probe design that is currently available involves exploiting the large data collection of known 16S rRNA sequences to design tailored probes for particular applications. There are few molecules within a cell that are as broadly conserved as rRNA. By developing a series of DNA probes directed towards rRNA (which is often present in microbes in upwards of 10 000 copies/cell), it is possible to choose probes that can detect rRNA sequences which are present only within a given sub-species, species, genus, or wider grouping (Kohne *et al.*, 1986). Probes have now been developed for a wide (Tenover, 1988) and ever-increasing spectrum of bacteria, viruses, fungi and protozoa, and these probes can be used in a variety of identification and typing situations.

3.2.2 TYPES OF PROBE LABEL AND DETECTION SYSTEMS

The development of new and modified methods for labelling and detecting nucleic acid probes is one of the most rapidly developing areas of molecular biology. The essential step involves labelling the probe, either directly or indirectly. In the former case, a detectable label is attached by a covalent bond directly to the nucleic acid of the probe, either internally or as an end-label (Figure 3.1). In the indirect method (Figure 3.2), a modification group is attached to the probe. The modification group is then detected indirectly by means of a specific labelled binding protein.

Methods of labelling nucleic acid probes, either directly with a detectable molecule or indirectly via a modification group, can be divided into chemical, enzymic or synthetic procedures (Matthews and Kricka, 1988; Keller and Manak, 1989; Leary and Ruth, 1989). Only a limited range of functional groups on a length of nucleic acid are suitable for attachment of a label. The various strategies available have been reviewed in detail by Kricka (1992).

(a) Direct labelling systems

Table 3.1 lists some of the main substances used for direct labelling of nucleic acid probes. In general, probe assays which are based on labels that provide signal amplification (i.e. enzymes) can be expected to be more sensitive than assays based on labels that provide only a single 'signal' per label (e.g. fluorescein). In all of the direct systems, the signal-generating reporter group is linked directly and covalently with the nucleic acid probe. Thus the detection of nucleic acids in direct labelling systems involves only two steps: hybrid formation between the target and the labelled probe, and signal generation by the covalently-linked reporter group.

Table 3.1 Examples of direct labelling systems for nucleic acid probes

Label	Reference
^{32}P	Southern (1975)
^{35}S	Collins and Hunsaker (1985)
Alkaline phosphatase	Renz and Kurz (1984)
Ethidium	Albarella and Anderson (1986)
Fluorescein	Amman *et al.* (1990)
Horseradish peroxidase	Urdea *et al.* (1988)
Microperoxidase	Heller and Shneider (1983)
Nitrobenzofuran	Draper (1984)
Tetramethylorhodamine	Amman *et al.* (1990)

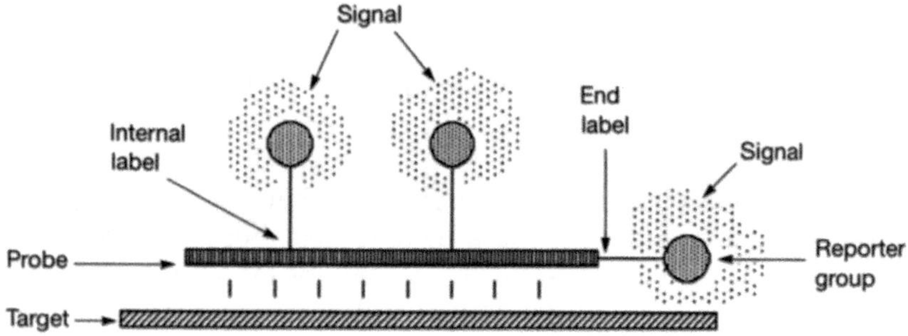

Figure 3.1 Schematic representation of direct labelling and detection method for nucleic acid probes.

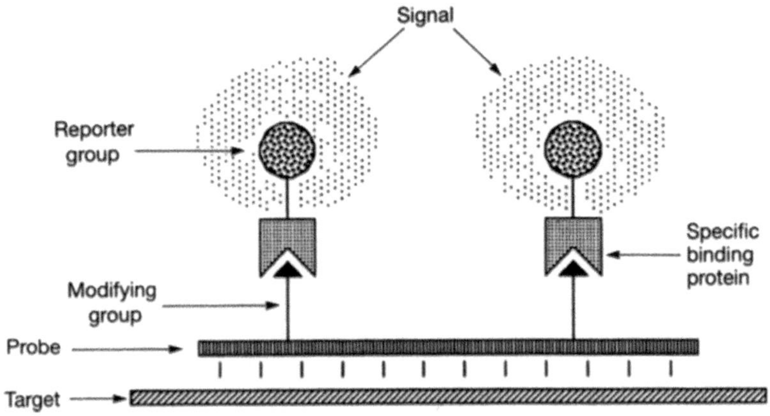

Figure 3.2 Schematic representation of indirect labelling and detection method for nucleic acid probes.

(b) Indirect labelling systems

In contrast to the direct systems, detection of nucleic acids in indirect labelling systems involves three reaction steps: hybrid formation between the target and the modified probe; a specific non-covalent reaction between the modified probe and a binding protein coupled with a reporter group; and signal generation by the reporter group. Examples of some of the more common indirect labelling systems are listed in Table 3.2.

Table 3.2 Examples of indirect labelling systems for nucleic acid probes

Modifying group/binding protein	Reporter group	Reference
Biotin–dUTP/streptavidin	Alkaline phosphatase	Leary *et al.* (1982)
Digoxigenin–dUTP/ antidigoxigenin	Alkaline phosphatase	Kessler *et al.* (1990)
Sulphone/antisulphone	Europium chelate	Syvänen *et al.* (1986)
Dinitrophenyl/ antidinitrophenyl	Pyruvate kinase	Keller *et al.* (1990)
Poly(dA)/poly(dT)	Horseradish peroxidase	Morrisey and Collins (1989)
Acetylaminofluorene/anti-acetylaminofluorene	Alkaline phosphatase	Tchen *et al.* (1984)

The most sensitive and popular indirect labelling systems are those based on biotin and digoxigenin. Both of these systems are capable of detecting sub-picogram amounts of nucleic acid, but the ubiquity of biotin in biological samples means that higher background reactions may be a problem under certain circumstances. The digoxigenin system avoids this problem since digoxigenin is a cardenolide that occurs only in *Digitalis* plants.

(c) Radioactive versus non-radioactive probe labels

Traditionally, hybridization assays have utilized radioactive probe labels such as ^{32}P and ^{35}S. After hybridization, binding of labelled probe DNA to the target is detected by autoradiography with X-ray film or by scintillation counting. Radioactive labels originally had some of the highest sensitivities amongst labelling methods, and are capable of detecting $\leqslant 1$ pg of target DNA. Unfortunately, radioactive labels are not suitable for routine diagnostic laboratories involved in microbial identification and typing. The main limitations include the relatively short half-lives of such labels (which restrict the shelf-life of labelled probes and kits containing them), and the possible health and safety hazards resulting from the use and disposal of radioactivity.

As a consequence of these limitations, non-radioactive labels have been the focus of much development in recent years, resulting in the availability of systems with sensitivities that now approach or equal those involving radioactive labels. Table 3.3 lists the properties of an ideal label for a nucleic acid probe. None of the non-radioactive labels listed in Tables 3.1 and 3.2 fulfils all of these criteria, although several go a considerable way towards them, and there is as yet no general agreement on the most appropriate non-radioactive label. However, the availability of sensitive detection methods has resulted in the wide use of enzyme-based systems, such as those incorporating alkaline peroxidase and horseradish peroxidase, in either direct or indirect labelling formats. Many such labelling systems were developed originally for use in immunoassays, but have now been applied to nucleic acid hybridizations. As a consequence, the use of radioactive labels is now confined almost entirely to specialized research laboratories.

Table 3.3 Properties of an ideal label for a nucleic acid probe

1. Easy to attach to nucleic acid by a simple and reproducible method
2. Stable under hybridization conditions and during storage
3. No interference with hybridization reaction
4. Applicable to solid phase or liquid phase hybridization reactions
5. Simple and sensitive detection procedure
6. Easy to remove for successive reprobing experiments
7. Distinguishable in the hybridized form from unhybridized probe in an homogeneous (non-separation) format
8. Applicable to automated analysis systems

(d) Non-radioactive detection strategies

The three most widely used strategies for detecting non-radioactive probe labels are as follows:

(i) Chemiluminescence and bioluminescence.

Both chemiluminescence and bioluminescence involve the emission and detection of light. Chemiluminescence is the emission of light that occurs because of decay of certain chemi-excited molecules to the electronic ground state. Novel chemiluminescent substrates, such as AMPPD (Bronstein, Edwards and Voyta, 1989), can be used as direct substrates for alkaline phosphatase in hybridization assays, with extremely sensitive limits of detection within a short period of time (Bronstein *et al.*, 1990). Bioluminescent detection of alkaline phosphatase activity is based on the release of D-luciferin from substrates such as firefly D-luciferin-*O*-phosphate (Miska and Geiger, 1987). The released D-luciferin then emits light following the addition of firefly luciferase. The original substrate, D-luciferin-*O*-phosphate, does not serve as a substrate for luciferase. Both chemiluminescence and bioluminescence are very sensitive, rapid and versatile detection methods.

(ii) Colorimetry.

Colorimetric assays produce either soluble or insoluble coloured products, but are relatively insensitive compared to some luminescent methods. Numerous chromogenic substrates have been developed for detection of enzymes such as alkaline phosphatase. They have the advantage of producing a simple visual end-product, and can be introduced without difficulty into virtually any routine microbiology laboratory.

One method of increasing the sensitivity of colorimetric systems is to couple them with an enzymic cascade reaction. A typical example might involve dephosphorylation of NADP to NAD by alkaline phosphatase associated with a nucleic acid probe assay. In such a system the NAD cofactor could then activate a cyclic secondary redox system involving alcohol dehydrogenase and diaphorase. In each cycle the NAD would be reduced to NADPH + H^+. This redox reaction could then be coupled to a secondary redox reaction in which NADPH + H^+ is oxidized to NAD, linked with the reduction of *p*-iodonitro-tetrazolium purple to formazan. The formazan could then be quantified photometrically (Self, 1985).

(iii) Fluorescence and time-resolved fluorescence.

Detection systems based on fluorescence are capable, potentially, of detecting single molecules of fluorescein (Mathies and Stryer, 1986). In practice, high background signals are obtained from the fluorescence which is present in all biological samples. This problem can be overcome by using a long-lived fluorophore that is excited by a pulse of excitation light. The emission is then measured after the relatively short-lived background fluorescence has decayed (Soini and Lövgren, 1987; Lövgren and Pettersson, 1990). For detection of alkaline phosphatase, the time-resolved detection of a lanthanide chelate reagent is combined with the activity of the enzyme itself to give a method termed **enzyme-amplified lanthanide luminescence** (EALL; Evangelista, Pollack and Templeton, 1991). In this method, a substrate that is incapable of forming a luminescent lanthanide chelate is converted by alkaline phosphatase into a compound that does form such a chelate. The method is applicable to a wide variety of assay formats, and is extremely sensitive, but has the disadvantage of requiring specialized instrumentation to excite and resolve the fluorescence.

Many non-radioactive labelling and detection systems are now available commercially. A complete description of the various experimental or commercially available labelling systems is beyond the scope of this volume; interested readers are referred to Kricka (1992) for a full treatment of the subject.

3.2.3 HYBRIDIZATION FORMATS

Hybridization reactions for microbial identification and typing purposes can be performed in one of three major formats: *in situ*; in solution; or on a solid support. All three formats are available for use in identification tests, but only the solid-support format has so far been used for typing purposes.

(a) *In-situ* hybridization techniques

In-situ hybridizations have, until recently, been used primarily to localize nucleic acid sequences within tissue sections fixed on microscope slides (Matthews and Kricka, 1988). However, most microorganisms are permeable to short labelled oligonucleotide probes following fixation and, by utilizing probes labelled with specific fluorescent dyes, it is possible to visualize particular organisms directly by fluorescence microscopy (Giovannoni *et al.*, 1988; DeLong, Wickham and Pace, 1989; Amman *et al.*, 1989). Hybridization must be performed in a sealed chamber to prevent loss of the hybridization solution by evaporation; such loss results in non-specific binding of the fluorescent probe to the cells. The temperature of hybridization depends upon each specific oligonucleotide, and must be optimized empirically. Following hybridization, slides can be viewed immediately or stored in the dark for up to six months. If the probe is specific for an rRNA target, the sensitivity is increased greatly since there are generally many thousands of rRNA copies per cell. These techniques have been reviewed in detail by Stahl and Amman (1991).

(b) Solution hybridization reactions

Hybridization reactions in a solution phase occur more rapidly than with a target sequence bound to a solid phase. This is because both the target and probe molecules are free to move, thereby maximizing the possibility that hybridization between complementary nucleic acid sequences will occur successfully. In general, solution hybridization

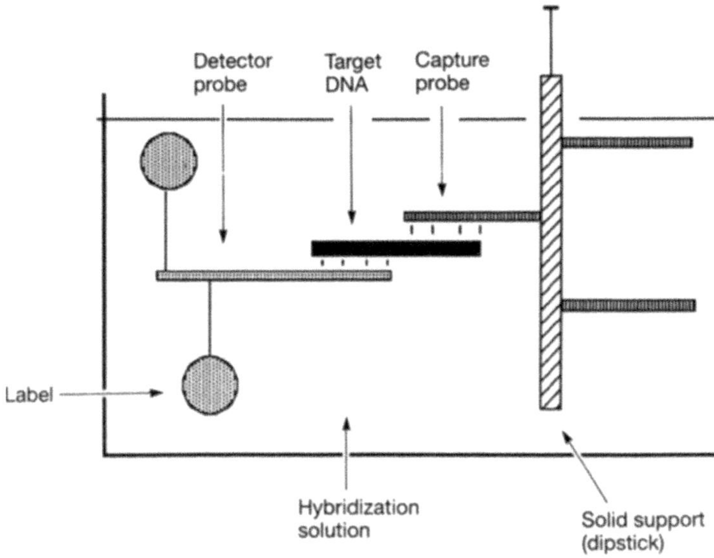

Figure 3.3 Example of a sandwich hybridization format for use in liquid phase hybridization reactions.

reactions will go to completion between five and ten-fold more rapidly than hybridization reactions on solid supports (Bryan *et al.*, 1986), and accelerators can be added to increase the rate of hybridization even further. An additional post-hybridization separation step is, however, required to detect probe-target hybrids. This drawback is countered by the advantages that, because the target sequences are kept in solution, there is little problem with background and less preparation of the sample is required.

Most commercially available hybridization kits for detecting specific microorganisms are based on an in-solution format. Microorganisms contained in samples are first lysed by a procedure appropriate to the organism being detected. Following hybridization, separation of probe-target hybrids is usually based on one of several variations of a '**sandwich hybridization**' format. Sandwich hybridizations require two adjoining nucleic acid sequences for use as 'capture' and 'detector' probes. It is important that there is no homology between these two sequences, although both should have homology with different regions of the target sequence. Labelled detector probe is added in solution to the target nucleic acid released from the sample being studied. After a short period for hybridization to occur, the capture probe immobilized on a solid support (e.g. a dipstick) is added to the reaction. The detector probe binds to the solid support via its hybridization to the target sequence (Figure 3.3). The label/detector–probe/target/capture–probe complex can then be removed to an appropriate substrate for detection of the label. This facilitates the automated reading of results if a colorimetric or fluorescent detection system is used.

A number of variations on the general approach outlined above are available (Wolcott, 1992). All in-solution hybridization reactions are suitable for direct and sensitive detection of individual specific microorganisms, but have not been utilized to date in typing applications.

(c) Membrane-based techniques

Immobilization of DNA on nitrocellulose membranes was first described by Nygaard and Hall (1963, 1964), but it was the description by Southern (1975) of the separation of DNA fragments by gel electrophoresis, their transfer to nitrocellulose filters, and the subsequent detection of specific sequences by hybridization that precipitated a major step forward in molecular biology. Since 1975, many modified transfer and hybridization protocols have been described (Meinkoth and Wahl, 1984), and many alternative supports, such as activated or strengthened nitrocellulose and nylon membranes, have been utilized.

For straightforward identification of a microorganism by hybridization on a solid support, it is sufficient usually to spot the sample (which can be purified nucleic acid, a suspension of a pure culture, or a fairly crude specimen of mixed biological material) onto a suitable membrane support. The sample is lysed and the nucleic acid is denatured, after which it is attached to the filter either by baking in a vacuum oven at 80°C for 2 h or by cross-linking (to nylon membranes) by means of exposure to ultraviolet light. The labelled probe is then added in solution following a prehybridization step to prevent non-specific binding of labelled probe to the filter.

When the hybridization step is completed, the filter is washed to remove unbound or partially-bound probe. It is this washing step which determines the stringency of the reaction. The labelled bound probe is then detected with an appropriate detection system. The whole process of hybridization of nucleic acids on solid supports has been reviewed in detail by Meinkoth and Wahl (1984).

For typing applications, it is necessary to utilize a technique based on that of Southern (1975). Purified chromosomal DNA is digested with appropriate restriction endonucleases and electrophoresed on an agarose gel to generate a chromosomal 'fingerprint' which, as discussed earlier, may often be of extreme complexity. The gel is then placed immediately underneath a nitrocellulose or nylon membrane sandwiched between several pieces of buffer-soaked filter paper. Paper

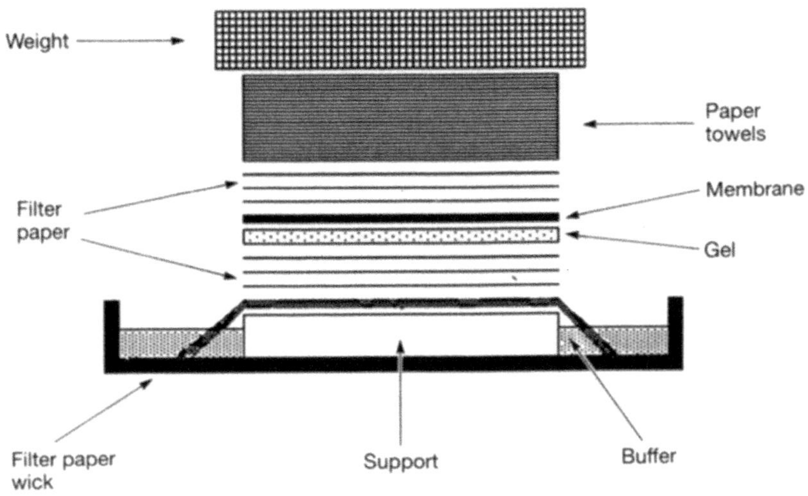

Figure 3.4 Schematic representation of a Southern blot.

towels are placed on top of the 'sandwich' (Figure 3.4), with the result that buffer is drawn by capillary action through the agarose gel. This causes the individual DNA fragments to be transferred in a flow of liquid from the gel to the nitrocellulose or nylon membrane, where they bind to form an exact copy or 'blot' of the original arrangement of fragments in the agarose gel (Sambrook, Fritsch and Maniatis, 1989). This 'traditional' method can be modified for nylon membranes by performing the transfer under denaturing conditions (Reed and Mann, 1985). After several hours, often overnight, the DNA fragments are attached permanently to the filter by baking or cross-linking as described earlier. The RFLPs can then be analysed by hybridization with an appropriate probe (section 3.3).

Alternative, more rapid, active methods of transferring nucleic acid fragments from agarose gels to nitrocellulose or nylon filters have also been devised (Sambrook *et al.*, 1989). These methods include '**electroblotting**' (Bittner, Kupferer and Morris, 1980) and '**vacuum blotting**' (Medveczky *et al.*, 1987; Olszewska and Jones, 1988).

For electroblotting, the DNA fragments contained in the agarose gel are first depurinated, denatured and neutralized *in situ*. The gel is then sandwiched between two slab electrodes next to a blotting membrane (Figure 3.5). When an electric field is applied, the nucleic acid molecules migrate from the gel to the membrane, where they are bound as described previously. The time required for complete transfer depends on the size of the DNA fragments, but is generally complete within 1–3 h. Electroblotting can be performed in either a vertical or a horizontal

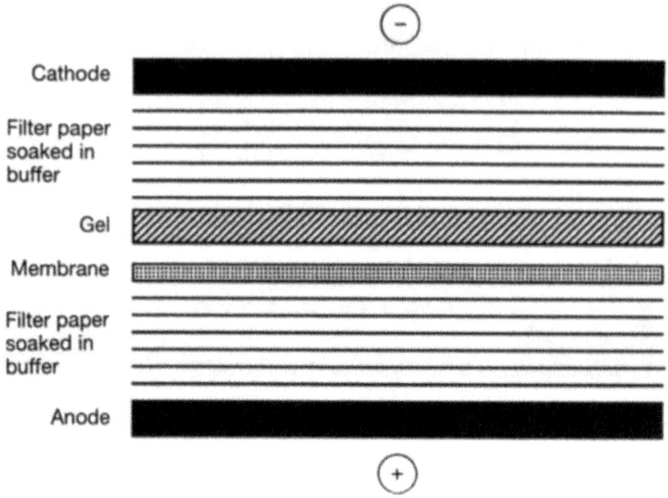

Figure 3.5 Schematic representation of an electroblot.

format, and several different designs of apparatus are available commercially. With the vertical format, the blotting 'sandwich' is held vertically and is submerged in buffer. Fairly high currents (with consequent raised temperatures) and large volumes of buffer are required. In contrast, the horizontal format uses a 'semi-dry' system in which the gel and membrane are sandwiched between layers of filter paper soaked in an appropriate buffer. For semi-dry blotting the consumption of buffer is negligible, and a current of only 1 mA/cm^2 is required.

For vacuum blotting, the gel is placed in contact with a nitrocellulose filter or nylon membrane supported over a vacuum chamber (Figure 3.6). Application of a low-pressure vacuum results in nucleic acid fragments being eluted from the gel and deposited on the membrane. Again, several different types of apparatus are available commercially. Vacuum blotting is extremely rapid and is thought to be considerably more efficient than capillary transfer. Optimal transfer of DNA fragments from genomic digests can be obtained in less than 1 h.

3.3 ANALYSIS OF RFLPs BY HYBRIDIZATION TECHNIQUES

3.3.1 BASIC CONCEPTS AND APPLICATIONS

As outlined earlier, the complexity of some RFLP patterns can make it extremely difficult to obtain useful typing information for epidemiological purposes. It is particularly difficult to compare the results obtained

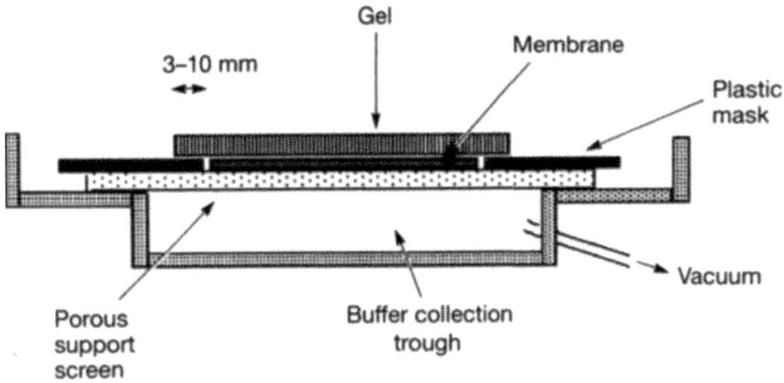

Figure 3.6 Schematic representation of vacuum blot.

on different gels or in different laboratories, since the fragments can be so numerous that it is impossible to read their sizes accurately and reproducibly. However, if the complex pattern of restriction fragments is denatured and transferred by blotting techniques from the gel to a nitrocellulose or nylon membrane and then hybridized with a suitable labelled probe, a much simpler pattern can be visualized which consists only of fragments that generate a positive hybridization signal with the probe. The relatively small number of such fragments means that their sizes can be determined much more precisely and recorded for inter-gel and inter-laboratory comparisons.

Several different types of probe can be used (Grimont and Grimont, 1991):

1. The probe can consist of ribosomal RNA (rRNA) from a single species. rRNA from *Escherichia coli* is available commercially (Boehringer Mannheim) and forms a convenient probe which can be labelled readily with either radioactive or non-radioactive reporter molecules. A major advantage of this approach is that, because of the conserved nature of parts of rRNA sequences, such a probe is a 'universal' probe which is able to hybridize with different DNA restriction fragments from a wide variety of bacterial species (Grimont and Grimont, 1986; Stull, LiPuma and Edlind, 1988). Some species are so homogeneous that all strains examined with a probe consisting of labelled rRNA will be found to give a single rRNA hybridization pattern, while different strains in other species give widely diverse patterns. When many different patterns are observed within a species, the method forms the basis of 'ribotyping' (section 3.3.2).
2. The probe can consist of a randomly cloned DNA fragment of unknown function (Tompkins *et al.*, 1986). Such a probe is often applicable only to the species from which it was derived originally,

although the method itself is a general one. Use of such a probe can sometimes be helpful in subdividing rRNA hybridization patterns (Saunders *et al.*, 1990; Tram *et al.*, 1990).

3. Alternatively, the probe can consist of a cloned fragment of a known gene. Thus, a fragment of the gene coding for exotoxin A has been used to type *Pseudomonas aeruginosa* (Ogle *et al.*, 1987), and a probe consisting of the cholera toxin gene has been used to type toxin-producing strains of *Vibrio cholerae* (Yam, Li Lung and Ng, 1989). Cloned fragments of rRNA genes can be used to give rRNA hybridization patterns which are similar, in concept, to those obtained with the labelled rRNA probes referred to above.

Use of any of these probe types can reduce greatly the number of bands in the original chromosomal fingerprint, and thereby facilitate comparisons between strains. The main disadvantage of this approach is that information is provided only about specific regions within the total genome that happen to hybridize with the particular probe being used. The technique of ribotyping is dealt with in detail in the following section. Examples of the analysis of RFLPs in microorganisms with other probe types are listed in Table 3.4

Table 3.4 Examples of the application of RFLP analysis with non-ribosomal probes to the typing of microorganisms

Organism	Reference
Aeromonas spp.	Altwegg and Luthyhottenstein (1991)
Borrelia burgdorferi	Wallich *et al.* (1992)
Candida albicans	Schmid *et al.* (1992)
Chlamydia trachomatis	Scieux *et al.* (1992)
Corynebacterium diphtheriae	Groman *et al.* (1983)
Cryptococcus neoformans	Spitzer and Spitzer (1992)
Escherichia coli	Böhm and Karch (1992)
Haemophilus influenzae	Forbes *et al.* (1992)
Histoplasma capsulatum	Keath *et al.* (1992)
Lactobacillus helveticus	Delosreyesgavilan *et al.* (1992)
Legionella pneumophila	Harrison *et al.* (1992)
Leptospira interrogans	Pacciarini *et al.* (1992)
Mycobacterium tuberculosis	Mazurek *et al.* (1991)
Mycoplasma spp.	Yogev *et al.* (1988b)
Pseudomonas aeruginosa	Loutit and Tompkins (1991)
Rhizobium leguminosarum	Demezas *et al.* (1991)
Rotavirus	Sethabutr *et al.* (1990)
Salmonella spp.	Tompkins *et al.* (1986)
Shigella spp.	Soldati and Piffaretti (1991)
Staphylococcus aureus	Goh *et al.* (1992)
Streptococcus pyogenes	Kaufhold *et al.* (1992)
Xanthomonas spp.	Lazo *et al.* (1987)

3.3.2 RIBOTYPING

Although each of the above probe types can generate highly satisfactory results, rRNA-based probes provide a system which is applicable to investigations of the molecular epidemiology of widely diverse bacteria, whereas the other probe types may be limited to particular species or strains within a species. The use of rRNA-based probes for typing purposes is referred to as '**ribotyping**'. Since its first description by Grimont and Grimont (1986), ribotyping has emerged as one of the most powerful of the currently available methods for investigating the epidemiology of microorganisms on a molecular basis.

The rationale for using rRNA genes as a basis for fingerprinting purposes has its origin in the high degree of conservation found in certain regions of such nucleic acid sequences (Woese, 1987). Many of the rRNA sequences found in bacteria appear to have changed little during evolution, and so probes specific for these sequences can detect a wide range of bacteria with similar rRNA sequences.

rRNA genes in bacteria are organized into operons, within which the individual genes coding for 16S, 23S and 5S RNA are often separated by non-coding spacer DNA. Use as a probe of a labelled mixture of 16S and 23S RNA results in hybridization only with the fragments in the chromosomal fingerprint which contain parts of the corresponding genes. In contrast, use as a probe of cloned fragments of the rRNA genes themselves may result in hybridization with the corresponding parts of the rRNA genes **and** the spacer sequences. Thus, different hybridization patterns can be obtained depending on the precise probe used (Saunders *et al.*, 1990).

(a) Technical considerations

The use of ribotyping for studying the molecular epidemiology of microorganisms requires that a certain amount of consideration be given to each stage in the procedure in order that optimal results are obtained (Grimont and Grimont, 1991).

First, it is necessary to generate a suitable restriction endonuclease fingerprint. An ideal fingerprint will consist of well resolved fragments with an even distribution of sizes. It is important not to have too many overlapping fragments. The number of fragments obtained will vary according to the source of DNA and the precise restriction endonuclease used. Accordingly, it is often necessary to test a few strains of a particular species with a range of enzymes before selecting the precise enzyme (or combination of enzymes) to be used for typing purposes. It is worth nothing that enzymes which recognize pentanucleotide restriction sites will usually generate more fragments than enzymes which recognize hexanucleotide restriction sites.

Even when a suitable enzyme (or combination of enzymes) to generate an acceptable fingerprint has been identified, the eventual data generated by ribotyping following digestion of chromosomal DNA with different restriction endonucleases may vary in its usefulness. It is therefore important to perform a thorough screening with well defined strains and several different restriction endonucleases before arriving at definite conclusions in epidemiological investigations. Results of studies that have relied on the use of only one or two enzymes may be misleading.

Once an acceptable fingerprint has been generated, it is then necessary to transfer the fragments to a blotting membrane. Of the various techniques available (section 3.2.3c), it seems that vacuum transfer may be the method of choice for resolution of close DNA fragments since it results in thinner bands than the other transfer methods. So far as the hybridization process itself is concerned, it is important to note that if labelled rRNA from a single species is used as a probe, it may be necessary to adjust the hybridization conditions, particularly the temperature, in order to permit successful hybridization with DNA fragments from a wide variety of microorganisms.

Various labels have been used in conjunction with rRNA probes. Isotopic labels such as ^{32}P (Grimont and Grimont, 1986; Stull *et al.*, 1988) have the advantage that relatively few steps are required for hybridization and washing, but all the disadvantages that are associated with the use of radioactivity in routine microbiology laboratories, including exposure hazards, short half-lives and waste-disposal problems. Several non-isotopic methods have also been applied successfully to ribotyping. Pitcher *et al.* (1987) described the synthesis of a biotinylated DNA probe to detect rRNA cistrons in *Providencia stuartii*, and enhanced chemiluminescence has been used to examine rRNA restriction patterns in *Vibrio cholerae* (Koblavi, Grimont and Grimont, 1990). rRNA has been labelled chemically with acetylaminofluorene (AAF) (Grimont *et al.*, 1989), and rRNA-AAF is available commercially (Eurogenetek, Belgium). More recently, Gustafero and Persing (1992) have described a novel procedure for direct covalent coupling of a horseradish peroxidase–polyethyleneimine complex to rRNA, followed by enhanced chemiluminescence. Probes carrying such non-isotopic labels are simple to synthesize and provide rapid results which are now comparable to those obtained with isotopic labels.

Finally, while it is relatively simple to compare rRNA hybridization patterns from different strains on a single membrane, accurate determination of fragment sizes is vital to establish an identification/typing system that can be shared between different laboratories. Sizing of

Table 3.5 Examples of microorganisms investigated by ribotyping

Organism	Reference
Acidovorax delafieldii	Aznar *et al.* (1992)
Acinetobacter spp.	Gerner-Smidt (1992)
Aeromonas spp.	Moyer *et al.* (1992)
Aspergillus spp.	Spreadbury *et al.* (1990)
Babesia spp.	Conrad *et al.* (1992)
Bacillus sphaericus	Demuro *et al.* (1992)
Bacteroides spp.	Smith and Callihan (1992)
Campylobacter jejuni	Wachsmuth *et al.* (1991)
Candida spp.	Magee *et al.* (1987)
Chlamydia trachomatis	Scieux *et al.* (1992)
Corynebacterium spp.	Soto *et al.* (1991)
Cyanobacteria	Nichols *et al.* (1982)
Enterobacter cloacae	Garaizar *et al.* (1991)
Enterococci	Lacoux *et al.* (1992)
Escherichia coli	Wachsmuth *et al.* (1991)
Gemmata spp.	Menke *et al.* (1991)
Haemophilus ducreyi	Sarafian *et al.* (1991)
Haemophilus influenzae	Irino *et al.* (1988)
Helicobacter pylori	Owen *et al.* (1992)
Legionella pneumophila	Schoonmaker *et al.* (1992)
Leptospira spp.	Perolat *et al.* (1990)
Listeria monocytogenes	Graves *et al.* (1991)
Mycobacterium spp.	Bercovier *et al.* (1986)
Mycoplasma spp.	Yogev *et al.* (1988a)
Neisseria meningitidis	Woods *et al.* (1992)
Pasteurella multocida	Zhao *et al.* (1992)
Planctomyces spp.	Menke *et al.* (1991)
Providencia stuartii	Owen *et al.* (1988)
Proteus vulgaris	Haun and Gobel (1987)
Pseudomonas aeruginosa	Hartmann *et al.* (1986)
Pseudomonas cepacia	Anderson *et al.* (1991)
Rhodococcus spp.	Lasker *et al.* (1992)
Salmonella typhi	Nastasi *et al.* (1991)
Serratia marcescens	Bingen *et al.* (1992)
Staphylococcus aureus	Blumberg *et al.* (1992a)
Staphylococcus epidermidis	Izard *et al.* (1992)
Shigella flexneri	Faruque *et al.* (1992)
Shigella sonnei	Hinojosa-Ahumada *et al.* (1991)
Group B streptococci	Blumberg *et al.* (1992b)
Thermus spp.	Menke *et al.* (1991)
Treponema spp.	Jensen *et al.* (1992)
Vibrio cholerae	Koblavi *et al.* (1990)
Yersinia enterocolitica	Blumberg *et al.* (1991)

fragments is achieved by precise comparison with different known size standards, but large percentage errors in size determination can occur. Owen and Beck (1987) found the algorithm of Elder and Southern (1983) to be the best method for estimating molecular sizes of digest fragments, while Irino *et al.* (1988) found that errors could be reduced to 1% with fragment sizes of <9 kb and the algorithm of Schaeffer and Sederoff (1981). Errors were larger (3–5%) when fragment sizes were >9 kb.

(b) Applications of ribotyping

Ribotyping has several advantages over other molecular typing methods. In particular, the ribosomal genes are extremely stable because they are located on the chromosome. Most microorganisms contain multiple copies of the ribosomal operons, so that hybridization with an appropriate labelled probe generates an acceptable number of fragments that yield a positive hybridization result. The fact that commercially available rRNA from *E.coli* can be labelled and used as a 'universal' probe means that the technique of ribotyping has been applied to a wide and ever-increasing range of microorganisms (Table 3.5). Even when detailed knowledge of a particular group of microorganisms is lacking, it is possible to characterize and compare strains belonging to virtually any species (including 'new' or previously undescribed species). An example of ribotyping is shown in Figure 3.7.

The diversity of ribopatterns obtained with certain species means that it may sometimes be difficult to rely simply on direct visual comparisons to determine strain similarities and differences in large-scale epidemiological studies. Owen *et al.* (1992) investigated the use of a standardized computer-assisted method for comparing ribopatterns obtained with different strains of *Helicobacter pylori*. Similarly, Bialkowska-Hobranska *et al.* (1990) used a laser densitometer for computerized analysis of ribopatterns generated with different strains of coagulase-negative staphylococci. Such methods may provide a means of determining similarities between different strains on an objective basis. The compilation of profile databases for different organisms may also allow the appearance of 'new' isolates to be identified and monitored.

It is important to note that the discriminatory ability of ribotyping may depend on the precise probe and restriction enzymes being used (e.g. Saunders *et al.*, 1991). These variables should be standardized if it is desired that results accumulated over an extended time period should be compared.

3.4 SUMMARY

In general, nucleic acid hybridization has been shown to have the potential for becoming an extremely valuable tool for typing and epidemiological studies with a wide range of microorganisms. The principal advantages and disadvantages of the method are summarized in Table 3.6. The major advantage is that the use of broad-spectrum probes, such

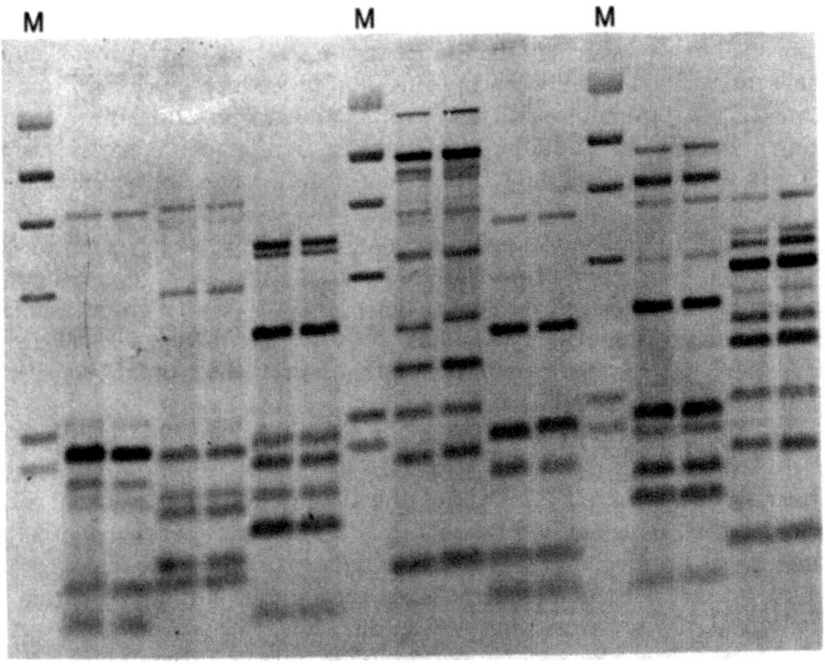

Figure 3.7 Example of the rRNA gene patterns (ribopatterns) for *Hae*III digests of chromosomal DNA from *Helicobacter pylori* probed with biotinylated cDNA from rRNA of strain NCTC 11638. Molecular size markers in lanes marked 'M' consist of Hin*d*III digests of bacteriophage lambda DNA. (Reproduced with the permission of Dr R.G. Owen and the publisher of the *Journal of Medical Microbiology*, **38**, 6–12.)

as those based on rRNA genes, offers a genuine 'universal' approach to identifying and/or typing any microorganism. However, several different probe types are available, and it is likely that specialized probes may provide better specificity or discrimination within particular species.

Table 3.6 Comparison of the main advantages and disadvantages of hybridization-based typing methods

Advantages
Applicable to a wide range of microorganisms
Commercially-available rRNA can be used as a 'universal' probe
Hybridization patterns are reproducible and relatively simple to interpret
Computer analysis can be used to compare patterns with information stored in databases
Disadvantages
Time-consuming and relatively complex methodology
Information is provided only about regions of the genome that hybridize with the particular probe being used

The availability of synthetic oligonucleotides for use as probes has improved many aspects of the hybridization strategy. Such probes eliminate the tedious isolation and purification procedures necessary to use cloned DNA fragments, and also avoid problems resulting from vector contamination. In addition, the hybridization reaction occurs more rapidly with oligonucleotide probes and has the potential for enhanced specificity.

Although hybridization analysis has now been used successfully to examine a wide range of microorganisms, in some cases (e.g. Schoonmaker, Heimberger and Birkhead, 1992), analysis of total chromosomal digest fingerprints by PFGE has proved useful for sub-dividing ribotypes and for distinguishing strains involved in outbreaks from epidemiologically unrelated strains.

Ribotyping and typing by PFGE are based on the same principle – the distribution of particular restriction enzyme sites on chromosomal DNA. While ribotyping examines the distribution of restriction sites within rRNA genes, or within the sequences surrounding this particular conserved chromosomal region, PFGE typing examines the distribution of particular restriction enzyme sites throughout the entire chromosome, and may therefore offer greater discriminative power with certain organisms. Thus, Prevost, Jaulhac and Piemont (1992) have shown that PFGE may be a more effective tool than ribotyping for the typing of methicillin-resistant strains of *Staphylococcus aureus*. The precise value of the different techniques will depend on the specific microorganism being studied. It is, therefore, important that hybridization patterns should be interpreted in conjunction with total DNA fingerprint patterns, since the latter may enable better discrimination between closely related strains in certain cases.

In conclusion, hybridization analysis of nucleic acid has proved to be a sensitive and specific tool for typing and epidemiological studies. The method is applicable immediately to investigations in specialized reference laboratories, even for 'unknown' microorganisms with no other available typing system, but procedures for nucleic acid isolation and purification, electrophoresis, blotting and hybridization are still labour-intensive, relatively complex and time-consuming. It will be necessary to simplify and accelerate these procedures before nucleic acid hybridization techniques are suitable for routine typing and epidemiological investigations in diagnostic laboratories. Given the current complexity of hybridization procedures, it is probable that hybridization analysis will remain for some time as an additional tool which can be used to assist in the typing of 'difficult' organisms or in the interpretation of differences identified by simpler, cheaper and more rapid typing methods.

REFERENCES

Albarella, J.P. and Anderson, L.H.D. (1986) Nucleic acid hybridization assay using antibodies to intercalating complexes with double-stranded nucleic acid. *U. S. Patent* **4563417**.

Altwegg, M. and Luthyhottenstein, J. (1991) Methods for the identification of DNA hybridization groups in the genus *Aeromonas*. *Experientia*, **47**, 403–6.

Amman, R., Krumholz, L. and Stahl, D. A. (1990) Fluorescent DNA probing of whole cells for determinative, phylogenetic and environmental studies in microbiology. *Journal of Bacteriology*, **172**, 762–70.

Anderson, D.J., Kuhns, J.S., Vasil, M.L. *et al.* (1991) DNA fingerprinting by pulsed field gel electrophoresis and ribotyping to distinguish *Pseudomonas cepacia* isolates from a nosocomial outbreak. *Journal of Clinical Microbiology*, **29**, 648–9.

Aznar, R., Owen, R.J. and Hernandez, J. (1992) DNA–DNA hybridization and ribotyping of *Acidovorax delafieldii* isolates from eels and aquatic environments. *Letters in Applied Microbiology*, **14**, 185–8.

Bercovier, H., Kafri, O. and Sala, S. (1986) Mycobacteria possess a surprisingly small number of ribosomal RNA genes in relation to the size of their genome. *Biochemical and Biophysical Research Communications*, **136**, 1136–41.

Bialkowska-Hobrzanska, H., Harry, V., Jaskot, D. and Hammerberg, O. (1990) Typing of coagulase-negative staphylococci by southern hybridization of chromosomal DNA fragments using a ribosomal RNA probe. *European Journal of Clinical Microbiology and Infectious Diseases*, **9**, 588–94.

Bingen, E.H., Marianikurkdjian, P., Lambertzechovsky, N.Y. *et al.* (1992) Ribotyping provides efficient differentiation of nosocomial *Serratia marcescens* isolates in a pediatric hospital. *Journal of Clinical Microbiology*, **30**, 2088–91.

Bittner, M., Kupferer, P. and Morris, C.F. (1980) Electrophoretic transfer of proteins and nucleic acids from slab gels to diazobenzyloxymethyl cellulose or nitrocellulose sheets. *Analytical Biochemistry*, **102**, 459–71.

Blanc, D.S., Siegrist, H.H., Sahli, R. and Francioli, P. (1993) Ribotyping of *Pseudomonas aeruginosa* – discriminatory power and usefulness as a tool for epidemiological studies. *Journal of Clinical Microbiology*, **31**, 71–2.

Blumberg, H.M., Kiehlbauch, J.A. and Wachsmuth, I.K. (1991) Molecular epidemiology of *Yersinia enterocolitica* O:3 infections – use of chromosomal DNA restriction fragment length polymorphisms of rRNA genes. *Journal of Clinical Microbiology*, **29**, 2368–74.

Blumberg, H.M., Rimland, D., Kiehlbauch, J.A. *et al.* (1992a) Epidemiologic typing of *Staphylococcus aureus* by DNA restriction fragment length polymorphisms of rRNA genes – elucidation of the clonal nature of a group of bacteriophage-nontypable, ciprofloxacin-resistant, methicillin-susceptible *S. aureus* isolates. *Journal of Clinical Microbiology*, **30**, 362–9.

Blumberg, H.M., Stephens, D.S., Licitra, C. *et al.* (1992b) Molecular epidemiology of group B streptococcal infections – use of restriction endonuclease analysis of chromosomal DNA and DNA restriction fragment length polymorphisms of ribosomal RNA genes (ribotyping). *Journal of Infectious Diseases*, **166**, 574–9.

Bronstein, I., Edwards, B. and Voyta, J.C. (1989) 1,2-Dioxetanes; novel chemiluminescent substrates: applications to immunoassays. *Journal of Bioluminescence and Chemiluminescence*, **4**, 99–111.

Bronstein, I., Voyta, J.C., Lazzari, K.G. *et al.* (1990) Rapid and sensitive detection of DNA in Southern blots with chemiluminescence. *BioTechniques*, **8**, 310–4.

Bryan, R.N., Ruth, J.L., Smith, R.D. and Le Bon, J.M. (1986) Diagnosis of clinical samples with synthetic oligonucleotide hybridization probes, in *Microbiology – 1986*, (ed. L. Leive), American Society for Microbiology, Washington, pp. 113–6.

Collins, M.L. and Hunsaker, W.R. (1985) Improved hybridization assays employing tailed oligonucleotide probes: a direct comparison with 5′-end-labeled oligonucleotide probes and nick-translated plasmid probes. *Analytical Biochemistry*, **151**, 211–24.

Conrad, P.A., Thomford, J.W., Marsh, A. *et al.* (1992) Ribosomal DNA probe for differentiation of *Babesia microti* and *B. gibsoni* isolates. *Journal of Clinical Microbiology*, **30**, 1210–15.

DeLong, E.F., Wickham, G.S. and Pace, N.R. (1989) Phylogenetic stains: ribosomal RNA-based probes for the identification of single cells. *Science*, **243**, 1360–3.

Delosreyesgavilan, C.G., Limsowtin, G.K.Y., Tailliez, P. *et al.* (1992) A *Lactobacillus helveticus* – specific DNA probe detects restriction fragment length polymorphisms in this species. *Applied and Environmental Microbiology*, **58**, 3429–32.

Demezas, D.H., Reardon, T.B., Watson, J.M. and Gibson, A.H. (1991) Genetic diversity among *Rhizobium leguminosarum* bv. Trifolii strains revealed by allozyme and restriction fragment length polymorphism analyses. *Applied and Environmental Microbiology*, **57**, 3489–95.

Demuro, M.A., Mitchell, W.J. and Priest, F.G. (1992) Differentiation of mosquito-pathogenic strains of *Bacillus sphaericus* from non-toxic varieties by ribosomal RNA gene restriction patterns. *Journal of General Microbiology*, **138**, 1159–66.

Draper, D.E. (1984) Attachment of reporter groups to specific, selected cytidine residues in RNA using a bisulfite-catalyzed transamination reaction. *Nucleic Acids Research*, **12**, 989–1001.

Elder, J.K. and Southern, E.M. (1988) Measurement of DNA length by gel electrophoresis II. comparison of methods for relating mobility to fragment length. *Analytical Biochemistry*, **128**, 227–31.

Evangelista, R.A., Pollack, A. and Templeton, E.F. (1991) Enzyme-amplified lanthanide luminescence for enzyme detection in bioanalytical assays. *Analytical Biochemistry*, **197**, 213–24.

Faruque, S.M., Haider, K., Rahman, M.M. *et al.* (1992) Differentiation of *Shigella flexneri* strains by rRNA gene restriction patterns. *Journal of Clinical Microbiology*, **30**, 2996–9.

Forbes, K.J., Bruce, K.D., Ball, A. and Pennington, T.H. (1992) Variation in length and sequence of porin (ompP2) alleles of noncapsulate *Haemophilus influenzae*. *Molecular Microbiology*, **6**, 2107–12.

Garaizar, J., Kaufmann, M.E. and Pitt, T.L. (1991) Comparison of ribotyping with conventional methods for the type identification of *Enterobacter cloacae*. *Journal of Clinical Microbiology*, **29**, 1303–7.

Gerner-Smidt, P. (1992) Ribotyping of the *Acinetobacter calcoaceticus-Acinetobacter baumannii* complex. *Journal of Clinical Microbiology*, **30**, 2680–85.

Giovannoni, S.J., DeLong, E.F., Olsen, G.J. and Pace, N.R. (1988) Phylogenetic group-specific oligodeoxynucleotide probes for identification of single microbial cells. *Journal of Bacteriology*, **170**, 720–6.

Goh, S.H., Byrne, S.K., Zhang, J.L. and Chow, A.W. (1992) Molecular typing of *Staphylococcus aureus* on the basis of coagulase gene polymorphisms. *Journal of Clinical Microbiology*, **30**, 1642–5.

Graves, L. M., Swaminathan, B., Reeves, M.W. and Wenger, J. (1991) Ribosomal DNA fingerprinting of *Listeria monocytogenes* using a digoxigenin-labeled DNA probe. *European Journal of Epidemiology*, **7**, 77–82.

Grimont, F. and Grimont, P.A.D. (1986) Ribosomal ribonucleic acid gene restriction patterns as potential taxonomic tools. *Annales de l' Institut Pasteur*, **137B**, 165–75.

Grimont, P.A.D. and Grimont, F. (1991) rRNA probes as tools for molecular epidemiology of microbial infections, in *Rapid Methods and Automation in Microbiology and Immunology*, (eds A. Vaheri, R.C. Tilton and A. Balows), Springer-Verlag, Berlin, pp. 47–53.

Grimont, F., Chevrier, D., Grimont, P.A.D. *et al.* (1989) Acetylaminofluorene-labelled ribosomal RNA for use in molecular epidemiology and taxonomy. *Research in Microbiology*, **140**, 447–54.

Groman, N., Cianciotto, N., Bjorn, M. and Rabin, M. (1985) Detection and expression of DNA homologous to the *tox* gene in nontoxigenic isolates of *Corynebacterium diphtheriae. Infection and Immunity*, **42**, 48–56.

Grothues, D. and Tümmler, B. (1991) New approaches in genome analysis by pulsed-field gel electrophoresis – application to the analysis of *Pseudomonas* species. *Molecular Microbiology*, **5**, 2763–76.

Gustaferro, C.A. and Persing, D.H. (1992) Chemiluminescent universal probe for bacterial ribotyping. *Journal of Clinical Microbiology*, **30**, 1039–41.

Harrison, T.G., Saunders, N.A., Haththotuwa, A. *et al.* (1992) Further evidence that genotypically closely related strains of *Legionella pneumophila* can express different serogroup specific antigens. *Journal of Medical Microbiology*, **37**, 155–61.

Haun, G. and Gobel, U. (1987) Oligonucleotide probes for genus-specific and subspecies-specific identification of representatives of the genus *Proteus*. *FEMS Microbiology Letters*, **43**, 187–93.

Heller, M.J. and Shneider, B.L. (1983) Chemiluminescent labeling of nucleic acids using azidophenyl glyoxal. *Federation Proceedings*, **42**, 1954.

Hinojosa-Ahumada, M., Swaminathan, B., Hunter, S.B. *et. al.* (1991) Restriction fragment length polymorphisms in rRNA operons for subtyping *Shigella sonnei*. *Journal of Clinical Microbiology*, **29**, 2380–4.

Irino, K., Grimont, F., Casin, I. *et al.* (1988) rRNA gene restriction patterns of *Haemophilus influenzae* biogroup *aegyptius* strains associated with Brazilian purpuric fever. *Journal of Clinical Microbiology*, **26**, 1535–8.

Izard, N.C., Hachler, H., Grehn, M. and Kayser, F.H. (1992) Ribotyping of coagulase-negative staphylococci with special emphasis on intraspecific typing of *Staphylococcus epidermidis*. *Journal of Clinical Microbiology*, **30**, 817–23.

Jensen, N.S., Casey, T.A., and Stanton, T.B. (1992) Characterization of *Serpulina*

(*Treponema*) *hyodysenteriae* and related intestinal spirochetes by ribosomal RNA gene restriction patterns. *FEMS Microbiology Letters*, **93**, 235–42.

Kaufhold, A., Podbielski, A., Johnson, D.R. *et al.* (1992) M-Protein gene typing of *Streptococcus pyogenes* by nonradioactively labeled oligonucleotide probes. *Journal of Clinical Microbiology*, **30**, 2391–7.

Keath, E.J., Kobayashi, G.S. and Medoff, G. (1992) Typing of *Histoplasma capsulatum* by restriction fragment length polymorphisms in a nuclear gene. *Journal of Clinical Microbiology*, **30**, 2104–7.

Keller, G.H. and Manak, M.M. (1989) *DNA Probes*. Stockton Press, New York.

Keller, G.H., Cumming, C.U., Huang, D.P. *et al.* (1988) A chemical method for introducing haptens onto DNA probes. *Analytical Biochemistry*, **170**, 441–50.

Kessler, C., Höltke, H.-J., Seibl, R. *et al.* (1990) Nonradioactive labeling and detection of nucleic acids: I. A novel DNA labeling and detection system based on digoxigenin:antidigoxigenin ELISA principle (digoxigenin system). *Molecular Genetics Hoppe-Seyler*, **371**, 917–27.

Koblavi, S., Grimont, F. and Grimont, P.A.D. (1990) Clonal diversity of *Vibrio cholerae* 01 evidenced by ribosomal RNA gene restriction patterns. *Research in Microbiology*, **141**, 645–57.

Kohne, D., Hogan, J., Jonas, V. *et al.* (1986) Novel approach for rapid and sensitive detection of microorganisms: DNA probes to rRNA, in *Microbiology – 1986*, (ed. L. Leive), American Society for Microbiology, Washington, pp. 110–2.

Kricka L.J., (ed). (1992) *Nonisotopic DNA Probe Techniques*. Academic Press, San Diego.

Lacoux, P.A., Jordens, J.Z., Fenton, C.M. *et al.* (1992) Characterizarion of enterococcal isolates by restriction enzyme analysis of genomic DNA. *Epidemiology and Infection*, **109**, 69–80.

Lasker, B.A., Brown, J.M. and McNeil, M.M. (1992) Indentification and epidemiological typing of clinical and environmental isolates of the genus *Rhodococcus* with use of a digoxigenin-labeled rDNA gene probe. *Clinical Infectious Diseases*, **15**, 223–33.

Lazo, G.R., Roffey, R. and Gabriel, D.W. (1987) Pathovars of *Xanthomonas campestris* are distinguishable by restriction fragment length polymorphism. *International Journal of Systematic Bacteriology*, **37**, 214–21.

Leary, J.J. and Ruth, J.L. (1989) Nonradioactive labeling of nucleic acid probes, in *Nucleic Acid and Monoclonal Antibody Probes*, (eds. B. Swaminathan and G. Prakash), Dekker, New York, pp. 33–57.

Leary, J.J., Brigati, D.J. and Ward, D.C. (1982) Colorimetric method for visualizing biotin-labeled DNA probes hybridized to DNA or RNA immobilized on nitrocellulose: Bio-blots. *Proceedings of the National Academy of Sciences of the USA*, **80**, 4045–9.

Loutit, J.S. and Tompkins, L.S. (1991) Restriction enzyme and Southern hybridization analyses of *Pseudomonas aeruginosa* strains from patients with cystic fibrosis. *Journal of Clinical Microbiology*, **29**, 2897–900.

Lövgren, T. and Pettersson, K. (1990) Time-resolved fluoroimmunoassay, advantages and limitations, in *Luminescence Immunoassay and Molecular Applications* (eds. K. Van Dyke and R. Van Dyke), CRC Press, Boca Raton, pp. 233–50.

Magee, B.B., D'Souza, T.M. and Magee, P.T. (1987) Strain and species identifi-

cation by restriction fragment length polymorphisms in the ribosomal DNA repeat of *Candida* species. *Journal of Bacteriology*, **169**, 1639–43.
Marconi, R.T., Lubke, L., Hauglum, W. and Garon, C.F. (1992) Species specific identification of and distinction between *Borrelia burgdorferi* genomic groups by using 16S rRNA-directed oligonucleotide probes. *Journal of Clinical Microbiology*, **30**, 628–32.
Mathies, R. A. and Stryer, L. (1986) Single molecule fluorescence detection: A feasibility study using phycoerythrin, in *Applications of Fluorescence in the Biomedical Sciences* (eds D. Lansing Taylor *et al.*), Liss, New York, pp. 129–40.
Matthews, J.A. and Kricka, L.J. (1988) Analytical strategies for the use of DNA probes. *Analytical Biochemistry*, **169**, 1–25.
Mazurek, G.H., Cave, M.D., Eisenach, K.D. *et al.* (1991) Chromosomal DNA fingerprint patterns produced with IS6110 as strain-specific markers for epidemiologic study of tuberculosis. *Journal of Clinical Microbiology*, **29**, 2030–3.
Medveczky, P., Chang, C.W., Oste, C. and Mulder, C. (1987) Rapid vacuum driven transfer of DNA and RNA from gels to solid supports. *BioTechniques*, **5**, 242.
Meinkoth, J. and Wahl, G. (1984) Hybridization of nucleic acids immobilized on solid supports. *Analytical Biochemistry*, **138**, 267–84.
Menke, M.A.O.H., Liesack, W. and Stackebrandt, E. (1991) Ribotyping of 16S and 23S ribosomal RNA genes and organization of *rrn* operons in members of the bacteria genera *Gemmata, Planctomyces, Thermotoga, Thermus* and *Verrucomicrobium*. *Archives of Microbiology*, **155**, 263–71.
Miska, W. and Geiger, R. (1987) I. Synthesis and characterization of luciferin derivatives for use in bioluminescence-enhanced enzyme immunoassays. New ultrasensitive detection systems for enzyme immunoassays. *Journal of Clinical Chemistry and Clinical Biochemistry*, **25**, 23–30.
Morrisey, D.V. and Collins, M.L. (1989) Nucleic acid hybridization assays employing dA-tailed capture probes. Single capture methods. *Molecular and Cellular Probes*, **3**, 189–207.
Moyer, N.P., Luccini, G.M., Holcomb, L.A. *et al.* (1992) Application of ribotyping for differentiating aeromonads isolated from clinical and environmental sources. *Applied and Environmental Microbiology*, **58**, 1940–4.
Nastasi, A., Mammina, C. and Villafrate, M.R. (1991) rDNA fingerprinting as a tool in epidemiological analysis of *Salmonella typhi* infections. *Epidemiology and Infection*, **107**, 565–76.
Nichols, J.M., Foulds, I.J., Crouch, D.H. and Carr, N.G. (1982) The diversity of cyanobacterial genomes with respect to ribosomal RNA cistrons. *Journal of General Microbiology*, **128**, 2739–46.
Nygaard, A.P. and Hall, B.D. (1963) A method for the detection of RNA-DNA complexes. *Biochemical and Biophysical Research Communications*, **12**, 98–104.
Nygaard, A.P. and Hall, B.D. (1964) Formation and properties of RNA-DNA complexes. *Journal of Molecular Biology*, **9**, 125–42.
Ogle, J.W., Janda, J.M., Woods, D.E. and Vasil, M.L. (1987) Characterization and use of a DNA probe as an epidemiological marker for *Pseudomonas aeruginosa*. *Journal of Infectious Diseases*, **155**, 119–26.
Olszewska, E. and Jones, K. (1988) Vacuum blotting enhances nucleic acid transfer. *Trends in Genetics*, **4**, 92–4.

Owen, R.J. and Beck, A. (1987) Evaluation of three procedures using a laser densitometer and microcomputer for estimating molecular sizes of restriction endonuclease digest fragments and application to *Campylobacter jejuni* chromosomal DNA. *Letters in Applied Microbiology*, **4**, 5–8.

Owen, R.J., Beck, A., Dayal. P.A., and Dawson, C. (1988) Detection of genomic variation in *Providencia stuartii* clinical isolates by analysis of DNA restriction fragment length polymorphisms containing rRNA cistrons. *Journal of Clinical Microbiology*, **26**, 2161–6.

Owen, R.J., Hunton, C., Bickley, J. *et al.* (1992) Ribosomal RNA gene restriction patterns of *Helicobacter pylori* – analysis and appraisal of *Hae*III digests as a molecular typing system. *Epidemiology and Infection*, **109**, 35–47.

Pacciarini, M.L., Savio, M.L., Tagliabue, S. and Rossi, C. (1992) Repetitive sequences cloned from *Leptospira interrogans* serovar Hardjo genotype Hardjoprajitno and their application to serovar identification. *Journal of Clinical Microbiology*, **30**, 1243–9.

Perolat, P., Grimont, F., Régnault, B. *et al.* (1990) rRNA gene restriction patterns of *Leptospira*: a molecular system. *Research in Microbiology*, **141**, 159–71.

Pitcher, D.G., Owen, R.J., Dyal, P. and Beck, A. (1987) Synthesis of a biotinylated DNA probe to detect ribosomal RNA cistrons in *Providencia stuartii*. *FEMS Microbiology Letters*, **48**, 283–7.

Prevost, G., Jaulhac, B. and Piemont, Y. (1992) DNA fingerprinting by pulsed-field gel electrophoresis is more effective than ribotyping in distinguishing among methicillin-resistant *Staphylococcus aureus* isolates. *Journal of Clinical Microbiology*, **30**, 967–73.

Reed, K.C., and Mann, D.A. (1985) Rapid transfer of DNA from agarose gels to nylon membranes. *Nucleic Acids Research*, **13**, 7207–21.

Renz, M. and Kurz, C. (1984) A colorimetric method for DNA hybridization. *Nucleic Acids Research*, **12**, 3435–44.

Sambrook, J., Fritsch, E.F. and Maniatis, T. (1989) *Molecular Cloning: A Laboratory Manual*, 2nd edn. Cold Spring Harbor Laboratory, New York.

Sarafian, S.K., Woods, T.C., Knapp, J.S. *et al.* (1991) Molecular characterization of *Haemophilus ducreyi* by ribosomal DNA fingerprinting. *Journal of Clinical Microbiology*, **29**, 1949–54.

Saunders, N.A., Harrison, T.G., Haththotuwa, A. *et al.* (1990) A method for typing strains of *Legionella pneumophila* serogroup 1 by analysis of restriction fragment length polymorphisms. *Journal of Medical Microbiology*, **31**, 45–55.

Saunders, N.A., Harrison, T.G., Haththotuwa, A. and Taylor, A.G. (1991) A comparison of probes for restriction fragment length polymorphism (RFLP) typing of *Legionella pneumophila*. *Journal of Medical Microbiology*, **35**, 152–8.

Schaeffer, H.E. and Sederoff, R.R. (1981) Improved estimation of DNA fragment lengths from agarose gels. *Analytical Biochemistry*, **115**, 113–22.

Schmid, J., Odds, F.C., Wiselka, M.J. *et al.* (1992) Genetic similarity and maintenance of *Candida albicans* strains from a group of AIDS patients, demonstrated by DNA fingerprinting. *Journal of Clinical Microbiology*, **30**, 935–41.

Schoonmaker, D., Heimberger, T. and Birkhead, G. (1992) Comparison of ribotyping and restriction enzyme analysis using pulsed-field gel electrophoresis for distinguishing *Legionella pneumophila* isolates obtained during a nosocomial outbreak. *Journal of Clinical Microbiology*, **30**, 1491–8.

Scieux, C., Grimont, F., Regnault, B. and Grimont, P.A.D. (1992) DNA fingerprinting of *Chlamydia trachomatis* by use of ribosomal RNA, oligonucleotide and randomly cloned DNA probes. *Research in Microbiology*, **143**, 755–65.

Self, C.H. (1985) Enzyme amplification – a general method applied to provide an immunoassisted assay for placental alkaline phosphatase. *Journal of Immunological Methods*, **76**, 389–93.

Sethabutr, O., Unicomb, L.E., Holmes, I.H. *et al.* (1990) Serotyping of human group A rotavirus with oligonucleotide probes. *Journal of Infectious Diseases*, **162**, 368–72.

Smith, C.J. and Callihan, D.R. (1992) Analysis of rRNA restriction fragment length polymorphisms from *Bacteroides* spp. and *Bacteroides fragilis* isolates associated with diarrhoea in humans and animals. *Journal of Clinical Microbiology*, **30**, 806–12.

Soini, E. and Lövgren, T. (1987) Time-resolved fluorescence of lanthanide probes and applications in biotechnology. *CRC Critical Reviews of Analytical Chemistry*, **18**, 105–54.

Soldati, L. and Piffaretti, J.C. (1991) Molecular typing of *Shigella* strains using pulsed field gel electrophoresis and genome hybridization with insertion sequences. *Research in Microbiology*, **142**, 489–98.

Soto, A., Pitcher, D.G. and Soriano, F. (1991) A numerical analysis of ribosomal RNA gene patterns for typing clinical isolates of *Corynebacterium* group D2. *Epidemiology and Infection*, **107**, 263–72.

Southern, E.M. (1975) Detection of specific sequences among DNA fragments separated by gel electrophoresis. *Journal of Molecular Biology*, **98**, 503–17.

Spitzer, E.D. and Spitzer, S.G. (1992) Use of a dispersed repetitive DNA element to distinguish clinical isolates of *Cryptococcus neoformans*. *Journal of Clinical Microbiology*, **30**, 1094–7.

Spreadbury, C.L., Bainbridge, B.W. and Cohen, J. (1990) Restriction fragment length polymorphisms in isolates of *Aspergillus fumigatus* probed with part of the intergenic spacer region from the ribosomal RNA gene complex of *Aspergillus nidulans*. *Journal of General Microbiology*, **136**, 1991–4.

Stahl, D. A. and Amann, R. (1991) Development and application of nucleic acid probes, in *Nucleic Acid Techniques in Bacterial Systematics* (eds E. Stackebrandt and M. Goodfellow), Wiley, Chichester, pp. 205–48.

Stull, T.L., LiPuma, J.J. and Edlind, T.D. (1988) A broad-spectrum probe for molecular epidemiology of bacteria: ribosomal RNA. *Journal of Infectious Diseases*, **157**, 280–6.

Syvänen, A.-C., Tchen, P., Ranki, M. and Soederlund, H. (1986) Time resolved fluorimetry: A sensitive method to quantify DNA-hybrids. *Nucleic Acids Research*, **14**, 1017–28.

Tchen, P., Fuchs, R.P.P., Sage, E. and Leng, M. (1984) Chemically modified nucleic acids as immunodetectable probes in hybridization experiments. *Proceedings of the National Academy of Sciences of the USA*, **81**, 3466–70.

Tenover, F.C. (1988) Diagnostic deoxyribonucleic acid probes for infectious diseases. *Clinical Microbiology Reviews*, **1**, 82–101.

Tompkins, L.S., Troup, N., Labigne-Roussel, A. and Cohen, M.L. (1986) Cloned, random chromosomal sequences as probes to identify *Salmonella* species. *Journal of Infectious Diseases*, **154**, 156–62.

Tram, C., Simonet, M., Nicolas, M.-H. *et al.* (1990) Molecular typing of nosocomial isolates of *Legionella pneumophila* serogroup 3. *Journal of Clinical Microbiology*, **28**, 242–5.

Urdea, M.S., Warner, B.D., Running, J.A. *et al.* (1988) A comparison of non-radioisotopic hybridization assay methods using fluorescent, chemiluminescent, and enzyme labeled synthetic oligodeoxyribonucleotide probes. *Nucleic Acids Research*, **16**, 4937–56.

Wachsmuth, I.K., Kiehlbauch, J.A., Bopp, C.A. *et al.* (1991) The use of plasmid profiles and nucleic acid probes in epidemiologic investigations of food-borne, diarrheal diseases. *International Journal of Food Microbiology*, **12**, 77–90.

Wallace, R.B., Shaffer, J., Murphy, R.F. *et al.* (1979) Hybridization of synthetic oligodeoxyribonucleotides to øX174 DNA. The effects of single base pair mismatch. *Nucleic Acids Research*, **6**, 3543–57.

Wallich, R., Helmes, C., Schaible, U.E. *et al.* (1992) Evaluation of genetic divergence among *Borrelia burgdorferi* isolates by use of *Ospa, Fla, Hsp60,* and *Hsp70* gene probes. *Infection and Immunity*, **60**, 4856–66.

Woese, C.R. (1987) Bacterial evolution. *Microbiological Reviews*, **51**, 221–71.

Wolcott, M.J., (1992) Advances in nucleic acid-based detection methods. *Clinical Microbiology Reviews*, **5**, 370–86.

Woods, T.C., Helsel, L.O., Swaminathan, B. *et al.* (1992) Characterization of *Neisseria meningitidis* serogroup C by multilocus enzyme electrophoresis and ribosomal DNA restriction profiles (ribotyping). *Journal of Clinical Microbiology*, **30**, 132–7.

Yam, W.C., Li Lung, M. and Ng, M.H. (1989) Molecular epidemiology of *Vibrio cholerae* in Hong Kong. *Journal of Clinical Microbiology*, **27**, 1900–2.

Yogev, D., Halachmi, D., Kenny, G.E. and Razin, S. (1988a) Distinction of species and strains of mycoplasmas (Mollicutes) by genomic DNA fingerprints with an rRNA probe. *Journal of Clinical Microbiology*, **26**, 1198–201.

Yogev. D., Sela, S., Bercovier, H. and Razin, S. (1988b) Elongation factor (EF-Tu) gene probe detects polymorphism in *Mycoplasma* strains. *FEMS Microbiology Letters*, **50**, 145–9.

Zhao, G.S., Pijoan, C., Murtaugh, M.P. and Molitor, T.W. (1992) Use of restriction endonuclease analysis and ribotyping to study epidemiology of *Pasteurella multocida* in closed swine herds. *Infection and Immunity*, **60**, 1401–5.

Nucleic acid amplification and sequencing techniques

4

4.1 INTRODUCTION

While nucleic acid hybridization procedures have now developed to the extent that they can be used readily for identification and typing of a wide range of microorganisms, the target nucleic acid is available only in very limited quantities unless the organism is first cultivated. Culture methods have not, however, been devised for some microorganisms, particularly viruses, while others may require a long period of cultivation to provide sufficient nucleic acid for detection or typing by hybridization techniques. This problem can be overcome by either artificially increasing the amount of target nucleic acid available for detection, or by using a signal amplification system that will allow a detectable signal to be generated from a minute amount of bound probe in a hybridization reaction. Various strategies for achieving such amplifications have been developed, and these are considered in this chapter.

Application of nucleic acid amplification methods may allow a million-fold or more amplification of a particular DNA or RNA sequence of interest. Various identification and typing methods can then be used for analysis of the amplification products. The simplest method is to use gel electrophoresis to identify the size of the reaction products, while better specificity and sensitivity can be obtained by analysing the reaction products by hybridization with a specific probe. The most informative method, however, involves determining the nucleotide sequence of the amplified DNA or RNA molecule. Identification of variations in nucleotide sequence allows the most precise identification and typing of microorganisms. While nucleic acid sequencing is not yet a technique for routine use, this chapter also outlines the main available sequencing techniques, and the ways in which they can be used to help establish evolutionary and/or epidemiological relationships among microorganisms.

4.2 THE POLYMERASE CHAIN REACTION (PCR)

The polymerase chain reaction (PCR) for amplification of specific nucleic acid sequences was introduced by Saiki *et al.* (1985), and has subsequently proved to be one of the most important scientific innovations of the past decade. (Methods for DNA amplification by the polymerase chain reaction are covered by patents held or pending by Hoffmann-La Roche Inc.)

PCR uses a thermostable polymerase to produce multiple copies of specific nucleic acid regions quickly and exponentially, including non-coding regions of DNA as well as particular genes. For example, starting with a single copy of a 1 kb DNA sequence, 10^{11} copies (or 100 ng) of the same sequence can be produced within a few hours. Once the reaction has occurred, a number of methods for identification and characterization of the amplification products are then applicable, of which the simplest is to identify the products according to their size following migration on agarose gels. For many diagnostic applications, the simple visualization of a PCR product of characteristic size is a significant outcome since it indicates the presence of the target DNA sequence in the original sample. Better specificity and sensitivity can be obtained either by examination of RFLPs in the amplified products or by hybridization analysis with specific probes, but the most informative and specific method is to determine the nucleotide sequence of the amplified DNA (section 4.4).

Following the introduction of PCR, it was recognized at an early stage that the method offered an exquisitely sensitive approach to the detection and identification of specific microorganisms in a variety of sample types. Potentially, a characteristic DNA sequence from a single virus particle or cell of a particular organism can be amplified to detectable levels within a very short period of time. The procedure can also be used to amplify RNA after a cDNA copy had been generated by reverse transcriptase (e.g. Wilson *et al.*, 1989). The method has received particular attention for detecting the presence of low numbers of bacteria or virus particles in clinical and environmental specimens. For example, detection of the HIV genome in infected blood may require finding HIV sequences present at one part per 100 000 human genome equivalents. While a diagnostic antibody response may take up to eight weeks to develop in an infected blood donor, specific HIV sequences can be detected in a few hours by PCR, either as an RNA molecule in

Table 4.1 Examples of the use of PCR for the specific identification of micro-organisms

Organism	Reference
Acanthamoeba spp.	Vodkin *et al.* (1992)
Bordetella pertussis	Houard *et al.* (1989)
Borrelia burgdorferi	Marconi and Garon (1992)
Brucella spp.	Herman and Deridder (1992)
Candida albicans	Miyakawa *et al.* (1992)
Carnobacterium spp.	Brooks *et al.* (1992)
Chlamydia trachomatis	Hayes *et al.* (1992)
Clostridium difficile	Gumerlock *et al.* (1991)
Coxiella burnetii	Stein and Raoult (1992)
Cryptosporidium parvum	Laxer *et al.* (1991)
Cytomegalovirus	Gozlan *et al.* (1992)
Dengue virus	Henchal *et al.* (1991)
Epstein–Barr virus	Samoszuk (1991)
Erwinia amylovora	Bereswill *et al.* (1992)
Escherichia coli	Jackson (1992)
Frankia spp.	Simonet *et al.* (1990)
Gaeumannomyces spp.	Henson (1992)
Helicobacter pylori	Clayton *et al.* (1992)
Hepatitis C virus	Okamoto *et al.* (1992)
Human immunodeficiency virus	Dawood *et al.* (1992)
Influenza virus	Claas *et al.* (1992)
Leptospira spp.	Merien *et al.* (1992)
Listeria monocytogenes	Niederhauser *et al.* (1992)
Luteovirus	Robertson *et al.* (1991)
Mycobacterium leprae	Hackel *et al.* (1990)
Mycobacterium tuberculosis	Altamirano *et al.* (1992)
Neisseria meningitidis	Maiden *et al.* (1992)
Papillomavirus	Charlotte *et al.* (1991)
Parvovirus	McOmish *et al.* (1993)
Plasmodium falciparum	Barker *et al.* (1992)
Pneumocystis carinii	Olsson *et al.* (1993)
Poliovirus	Yang *et al.* (1991)
Pseudomonas solanacearum	Seal *et al.* (1992a)
Rickettsia spp.	Gage *et al.* (1992)
Rotavirus	Taniguchi *et al.* (1992)
Salmonella spp.	Rahn *et al.* (1992)
Staphylococcus spp.	Murakami *et al.* (1991)
Toxoplasma gondii	Vandevan *et al.* (1991)
Treponema pallidum	Grimprel *et al.* (1991)
Trichomonas vaginalis	Riley *et al.* (1992)
Trypanosoma cruzi	Breniere *et al.* (1992)
Vibrio vulnificus	Hill *et al.* (1991)
Yersinia spp.	Nakajima *et al.* (1992)

an unintegrated virus particle or as a DNA molecule representing an integrated provirus (Murakawa *et al.*, 1988; Ou *et al.*, 1988).

The power of PCR has also been applied to the preparation of large quantities of specific DNA probes for use in hybridization reactions such as those described in the previous chapter. Probes can be labelled directly during the amplification reaction, either radioactively or non-radioactively (Schowalter and Sommer, 1989; An, Franklin and Fleming, 1992). Probes generated by PCR have a number of advantages over those labelled by more traditional methods such as nick-translation or random prime labelling. Only small amounts of template are required, short probes can be labelled extremely efficiently, specific probes from heterogeneous sequences can be generated, and rigorous purification of the template is not usually required.

Table 4.1 lists examples of the application of PCR to the specific detection of a range of microorganisms. Further examples can be found in almost every issue of scientific journals such as the *Journal of Clinical Microbiology* and *Applied and Environmental Microbiology*, and it would be futile to try and present an 'up-to-date' list here. Two recent books devoted entirely to PCR have been published (Erlich, 1989; Innis *et al.*, 1990), and the detailed information they contain will not be duplicated here. Rather, a brief outline of the basic methodology is given below, followed by a more detailed account of some specific uses of PCR and its more recently introduced variations for typing particular microorganisms.

4.2.1 BASIC PRINCIPLES OF PCR

(a) Outline of the methodology

PCR is an enzymic method for amplifying exponentially a specific pre-selected fragment of DNA. The amplification is achieved with a thermostable DNA polymerase, two synthetic oligonucleotide primers, and the four standard deoxyribonucleoside triphosphates that are incorporated into DNA.

Each PCR amplification is sub-divided into three steps which are repeated in cycles (Figure 4.1): (i) the 'melting' (strand separation) of the duplex sample DNA; (ii) the annealing of the two primers to opposite DNA strands; and (iii) the extension of the primers by polymerase-mediated nucleotide additions to produce two copies of the original sequence.

The specificity of the reaction is controlled by the oligonucleotide DNA strands and direct replication of the intervening region. In an exponential reaction, the original target sequence is then amplified a million-fold or more within a few hours.

The primers for PCR amplification are synthetic oligonucleotides which are 20–30 nucleotides long, and which are designed by reference to DNA sequence databases. Each pair of primers must act in concert

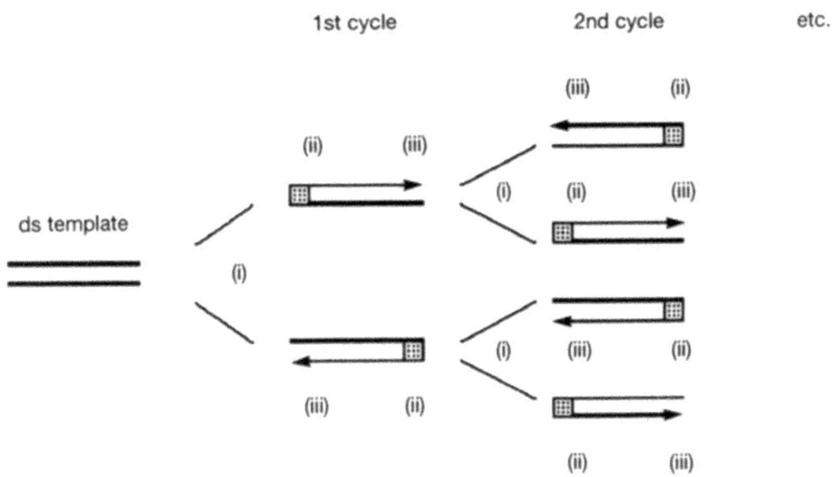

Figure 4.1 Schematic outline of the polymerase chain reaction (PCR). Each cycle consists of three steps: (i) thermal denaturation, typically at 94°C, to dissociate each DNA duplex; (ii) annealing of primers at a temperature determined by the required stringency of primer hybridization; (iii) elongation at an optimal temperature for thermostable polymerase activity, typically 72°C for *Taq* polymerase.

and must be relatively specific for their binding sites, since if one or both of the primers fails to hybridize, there will be no specific amplification products. With molecules that are highly conserved, such as rRNAs, it is possible to design amplification primers which will universally amplify all genes of a certain type. In contrast, if non-specific primers are used, non-specific amplification reactions may occur. Both of the above approaches may have applications for typing microorganisms (section 4.2.2).

The first PCR experiments used the Klenow fragment of *Escherichia coli* DNA polymerase I to extend the annealed primers. However, the Klenow fragment is denatured irreversibly at 94°C, and it was therefore necessary to add fresh enzyme after the denaturation step in each cycle. The process has now been improved greatly by the use of a thermostable DNA polymerase (*Taq*) isolated from *Thermus aquaticus* (Saiki *et al.*, 1988). *T. aquaticus* is a thermophilic eubacterium found in hot springs and water heaters. *Taq* DNA polymerase has a high optimal processing activity of 8000 bp/min at 75°C, but does not have a 3′-exonuclease activity (Innis *et al.*, 1988). The half-life of *Taq* polymerase in PCR reaction buffer is 130 min at 92.5°C, and 40 min at 95°C. Since *Taq* can, therefore, survive the PCR heat-denaturation step, fresh enzyme does not need to be added at each cycle. Although the repeated rapid shifts in temperature which are required for PCR amplification are difficult and tedious to achieve manually, a variety of automated thermal cyclers to perform this task are now available commercially.

(b) Technical considerations

Any PCR begins with the isolation of nucleic acid. While the amount of nucleic acid needed for a PCR is, theoretically, one copy of the target template, in practice 10–100 ng of DNA is often used. Crude preparations are adequate, but it is important to avoid the presence of large quantities of EDTA (which binds the free Mg^{2+} necessary for the PCR to proceed) and extraction buffers containing phosphate (which result in the precipitation of $Mg_3(PO_4)_2$ at high temperatures). Target sequences of up to 10 kb have been amplified (Jeffreys *et al.*, 1988), but since it is often easier to amplify shorter sequences, most applications choose a target length of <2 kb.

Table 4.2 Components of a typical PCR

DNA template	10–100 ng
dNTP (each nucleotide)	200 μM
Primer 1	0.1–1.0 μM
Primer 2	0.1–1.0 μM
Taq polymerase	1 unit
KCl	50 mM
Tris-HCl (pH 8.4 at 25°C)	10 mM
$MgCl_2$	2.5 mM
Gelatin	100 mg/l
Nonidet–40	0.05% (v/v)

Table 4.2 lists the typical components of a PCR. At the reaction temperatures optimal for *Taq* polymerase, it has been found that primers which are 20–30 nucleotides long provide a good balance between efficiency and specificity. Ideally, the composition of the primers should be as heterogeneous as possible, with an even mixture of all four bases and an avoidance of simple sequence motifs. Internal primer secondary structures (such as hairpin stems) should be absent, and the 3′ end of the primer should not have homology to internal regions of the product. Primer design is simplified considerably by the use of computer programs to check for internal and self-complementarities (Rychlik and Rhoads, 1989).

Non-specific amplification of DNA fragments is sometimes observed, usually following the use of degenerate amplification primers, but in most cases such non-specific amplification can be eliminated by careful adjustment of the reaction conditions. Temperature, Mg^{2+} concentration, and *Taq* polymerase concentration are all known to exert a strong effect on the specificity of a PCR. Inclusion of the non-ionic detergent Nonidet-40 in a reaction mixture has been reported to increase greatly the yield and reliability of PCR (Giovannoni, 1991).

Examination of RFLPs in PCR products offers a rapid means of discriminating between homologous genes. If amplified DNA is derived primarily from a single gene, then the PCR products can be digested directly with restriction endonucleases (see next section). However, before such detailed characterization of amplified DNA can proceed, it is essential to assess the purity of the amplified DNA. This can be done most conveniently by electrophoresis on an agarose gel, although the small size of most PCR products (generally <2 kb) means that an agarose formulation intended specifically for the separation of small DNA fragments should be used. Once the purity of the reaction products has been observed, the products are simply precipitated with ethanol, recovered by centrifugation, and then resuspended in a suitable buffer for the digestion reaction. Following digestion, the restriction fragments are visualized by electrophoresis on either an agarose or polyacrylamide gel. While agarose gels are often adequate for most purposes, higher resolution can be achieved with polyacrylamide gels.

Although conceptually quite simple, successful amplification by PCR involves balancing a large number of variables, including the quality and quantity of the starting template, the hybridization temperature, and the quality and activity of the *Taq* polymerase. A small 5% loss in efficiency in each amplification cycle will eventually result in a 50% reduction in amplification product after 25 cycles. Careful optimization of the various parameters involved in PCR is therefore a worthwhile exercise.

(c) Problems with PCR

PCR is an extremely powerful technique, but its very power can also lead to considerable problems, particularly when detecting virus or bacterial genes for diagnostic purposes. It is important to remember that nucleic acid from dead as well as viable microorganisms will give a positive reaction. Since even a single molecule of DNA can be amplified by PCR, it is also vital to prevent cross-contamination of DNA samples with amplified or foreign DNA. The slightest contamination of glassware, pipettes or reagents can result in the production of false-positive reactions. Such contamination problems impose a need for extreme cleanliness and rigorous controls. Amplification reactions should be performed in physical isolation (i.e. in a different room) from the parts of the laboratory where specimens are received and target nucleic acid is prepared. Various techniques for reducing extraneous

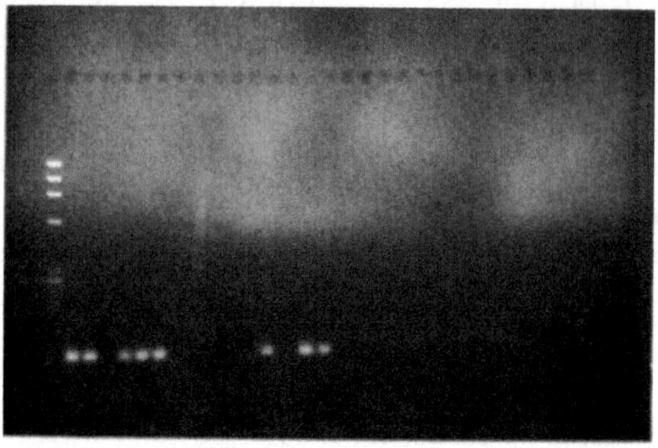

Figure 4.2 Example of the detection of hepatitis C virus DNA in human serum samples. Left hand track shows DNA size markers (ϕ174 bacteriophage DNA digested with *Hae*III). Remaining tracks show the 60 bp amplification product (part of the 5′ non-coding region of the hepatitis C virus genome) obtained following two rounds of nested PCR.

DNA contamination of PCR products have been described (see Wolcott, 1992), but it is vital that each set of PCR amplifications should include control reactions to verify the purity of reagents and the cleanliness of equipment.

Other problems may arise from the relatively high error rate of *Taq* polymerase. Base substitutions occur at about one in every 9000 bp, and frameshifts at about one in every 40 000 bp (Tindall and Kunkel, 1988). Although such error rates may seem to be insignificant, they may have profound effects on the homogeneity of the amplified products.

Although PCR can now be semi-automated, the technique still requires a certain amount of technical skill and some specialized equipment to prepare samples and perform amplification reactions successfully.

(d) Variations of PCR

A number of variations of the basic PCR technique have now been described, of which probably the most important for microbial identification and typing purposes are **nested PCR, reverse transcriptase PCR** and **asymmetric PCR**.

Nested PCR (Mullis and Faloona, 1987) involves two consecutive PCR reactions of about 25 cycles each. The first PCR reaction uses a pair of primers external to the sequence of interest. The second PCR uses the amplification products of the first reaction, in conjunction with one or more nested primers, to amplify a sequence derived from within the region flanked by the initial set of primers (Figure 4.2). The net result is a significant increase in the specificity and sensitivity of the assay (Porter-Jordan *et al.*, 1990).

Reverse-transcriptase PCR involves reverse transcription of an RNA template to produce RNA–cDNA hybrids, followed by PCR amplification, thereby effectively resulting in the amplification of an RNA target. This technique is particularly useful for detecting specific RNA from viruses (Jackson *et al.*, 1989), and can be combined with nested PCR to detect minute quantities of RNA (Garson *et al.*, 1990).

Asymmetric PCR (Gyllensten and Erlich, 1988) is a technique that facilitates the definitive identification of amplification products by enabling direct sequence analysis. This technique is described in more detail in section 4.4.2.

Many other variations of PCR have also been described (reviewed by Bej, Mahbuhani and Atlas, 1991; Erlich, Gelfand and Sninsky, 1991; Hayden *et al.*, 1991; Persing, 1991), but in many cases the protocols are not yet sufficiently developed and reproducible for routine use.

4.2.2. USES OF PCR FOR MICROBIAL TYPING PURPOSES

As indicated in the previous sections, specific PCR primer pairs can be used to identify genes that are characteristic of a particular species or even strain. Various strategies have also been devised by which the power of PCR can be harnessed for the epidemiological typing of microorganisms. These strategies can be divided conveniently into four major groups.

(a) Analysis of RFLPs generated by PCR with specific primers

PCR has been combined with restriction endonuclease digestion of amplified DNA to produce RFLPs for analysis on agarose or polyacrylamide gels. A typical example is the amplification of specific DNA sequences from the citrate synthase gene in rickettsiae, followed by subsequent cleavage of the amplified fragments with restriction endonucleases to differentiate different strains (Regnery, Spruill and Plikaytis, 1991). PCR coupled to restriction enzyme digestion requires only a fraction of the chromosomal DNA necessary for Southern blotting (Chapter 3) and can be completed in a much shorter time because of the elimination of the hybridization step. Examples of microorganisms where this type of approach has proved useful can be found in Table 4.3.

Table 4.3 Examples of microorganisms studied by analysis of RFLPs in specific non-ribosomal sequences amplified by PCR

Organism	Reference
Chlamydia spp.	Kaltenboek *et al.* (1992)
Clostridium spp.	Gurtler *et al.* (1991)
Comomonas spp.	van Eechoutte *et al.* (1992)
Cytomegalovirus	Sokol *et al.* (1992)
Helicobacter pylori	Foxall *et al.* (1992)
Hepatitis C virus	Okamoto *et al.* (1992)
Mycobacterium tuberculosis	Ross and Dwyer (1993)
Nitrobacter spp.	Navarro *et al.* (1992)
Papillomavirus	Lungu *et al.* (1992)
Rickettsia spp.	Regnery *et al.* (1991)
Streptococcus spp.	McClelland *et al.* (1992)

(b) PCR ribotyping

PCR ribotyping represents a specialized use of specific primers to amplify particular regions of the genome concerned with rRNA and tRNA (transfer RNA) synthesis. Traditional ribotyping (see Chapter 3) detects RFLPs by probing chromosomal DNA with rRNA. Although this is a powerful typing method with broad applicability for determining the molecular epidemiology of microorganisms, the technical skill and time required for the hybridization analysis are disadvantages which may limit its application. An alternative direct approach, which circumvents these problems, involves the use of PCR to detect polymorphisms in genes or intergenic spacer regions associated with rRNA or tRNA.

Most bacteria contain between 2 and 11 rRNA gene copies per bacterial cell (Gottlieb and Rudner, 1985; Srivastava and Schlessinger, 1990), while the intergenic spacer regions between the 16S and 23S rRNA genes encode various tRNAs and contain several direct repeat sequences in non-coding regions of the gene clusters (Bacot and Reeves, 1991). While the tRNA genes themselves are highly conserved, the lengths of the tRNA intergenic spacers are known to vary considerably, e.g. from 2 to 35 bp in *Bacillus subtilis* (Vold, 1985) and from 2 to 208 bp in *E. coli* (Jinks-Robertson and Nomura, 1987). The tandem arrangement of the tRNA gene clusters allows them to be explored for interspecific intergenic length polymorphisms with low stringency PCR employing primers derived from tRNA gene sequences. Using consensus sequence tRNA gene primers (Welsh and McClelland, 1991) it should be possible to generate suitable fingerprints from the majority of microorganisms. Table 4.4 lists examples of the few microorganisms where this approach has so far been used successfully. The banding patterns produced following PCR amplification of these RNA regions

Table 4.4 Examples of microorganisms fingerprinted by PCR ribotyping

Organism	Reference
Lactic acid bacteria	Klijn *et al.* (1991)
Listeria spp.	Czajka *et al.* (1993)
Mycoplasma spp.	van Kuppeveld *et al.* (1992)
Pseudomonas cepacia	Kostman *et al.* (1992)
Pseudomonas solanacearum	Seal *et al.* (1992b)
Staphylococcus spp.	Welsh and McClelland (1992)
Streptococcus spp.	McClelland *et al.* (1992)

seem to be reproducible and stable in a given isolate over time, and can therefore be used directly for typing purposes.

The high copy number of rRNA molecules in a cell means that sequence data of 16S rRNA genes can also be used to increase the sensitivity of identification by PCR. Engstrand *et al.* (1992) demonstrated that it was possible to use a specific oligonucleotide probe complementary to a 16S rRNA sequence as one of the primers in an RNA reverse transcription and cDNA PCR amplification assay (**RT-PCR**) for the detection of *Helicobacter* spp. The sensitivity of detection was increased up to 50-fold compared with conventional PCR assays.

(c) rep-PCR

An alternative method of generating fingerprints directly, i.e. without the use of restriction endonucleases, has been termed **repetitive element sequence-based PCR (rep-PCR)**.

The term rep-PCR refers to the general methodology involving the use of oligonucleotide primers based on short repetitive sequence elements that are dispersed throughout the prokaryotic kingdom. These elements appear to be conserved among many members of the Enterobacteriaceae and other bacterial species (Versalovic, Koeuth and Lupsky, 1991).

Consensus sequence probes matching two such repetitive elements, the 126 bp enterobacterial repetitive intergenic consensus (ERIC) sequence and the 38 bp repetitive extragenic palindromic (REP) sequence, both hybridize to genomic DNA from Enterobacteriaceae, other related Gram-negative bacteria, and a number of more distantly related bacterial species (Versalovic *et al.*, 1991). Outwardly directed primer sets based on the consensus sequences can be used to generate fingerprints following PCR amplification of genomic DNA from microorganisms which contain these sequences. Differences in band sizes following electrophoresis on agarose gels apparently represent polymorphisms in the distances between repetitive sequence elements in different genomes. The technique has been applied successfully, to date, in investigating relationships between a limited number of isolates of *Bacillus subtilis* (Versalovic *et al.*, 1992), *Citrobacter diversus* (Woods *et al.*, 1992) and *Rhizobium meliloti* (de Bruijn, 1992). The method generates quite complex fingerprints from most members of the Enterobacteriaceae (Versalovic *et al.*, 1991) and may be broadly applicable to fingerprinting all bacteria that possess such repetitive elements.

A similar method that allows the detection of variable DNA sequences in lower eukaryotes, such as the free-living amoeboflagellates belonging to the genus *Naegleria*, has also been described (van Belkum, de Jonckheere and Quint, 1992). This procedure involves the use of PCR primers targeted at sequences known to be polymorphic in both higher and lower eukaryotes (Jeffreys, Wilson and Thein, 1985) to generate individual fingerprints. As with bacteria, this technique should be broadly applicable to a range of lower eukaryotes.

(d) Randomly amplified polymorphic DNA fingerprints

While conventional PCR assays can be used to amplify DNA sequences that are characteristic of a particular species or even strain, the method has the drawback that specific oligonucleotide primers are required, thereby making knowledge of the DNA sequence of the organism being studied an essential prerequisite. Analysis of randomly amplified polymorphic DNA fingerprints (**RAPD analysis**), also known as **arbitrarily primed PCR** (**AP-PCR**), removes this requirement by using a primer(s) chosen without regard to the sequence of the genome to be fingerprinted (Welsh and McClelland, 1990; Williams *et al.*, 1990). Thus, this method requires no previous knowledge of the molecular biology of the organisms to be investigated, and can therefore be applied to any species from which DNA can be prepared.

The basis of the method is the observation by Welsh and McClelland (1990) that a single arbitrarily chosen primer, combined with two cycles of PCR at low stringency and many cycles at high stringency, generated a discrete and reproducible set of amplification products characteristic of particular genomes. The rationale for the phenomenon probably rests on the fact that, at a sufficiently low temperature, a primer can be expected to anneal to many sequences with a variety of mismatches. Some of these sequences will be within a few hundred bp of each other and on opposite strands, such that the intervening sequence will be capable of amplification by PCR. Arbitrary primers as short as 5 bp have been shown to generate complex genomic fingerprints following PCR amplification (Caetano-Annollés *et al.*, 1991). Table 4.5 lists examples of microorganisms fingerprinted by this procedure.

Table 4.5 Examples of microorganisms fingerprinted by RAPD analysis

Organism	Reference
Agaricus bisporus	Khush *et al.* (1992)
Aspergillus fumigatus	Aufauvre-Brown *et al.* (1992)
Bacillus thuringiensis	Brousseau *et al.* (1993)
Brucella spp.	Fekete *et al.* (1992)
Campylobacter spp.	Mazurier *et al.* (1992)
Candida spp.	Lehmann *et al.* (1992)
Casuarina spp.	Sellstedt *et al.* (1992)
Clostridium difficile	McMillan and Muldrow (1992)
Frankia spp.	Sellstedt *et al.* (1992)
Fusarium solani	Crowhurst *et al.* (1991)
Helicobacter pylori	Akopyanz *et al.* (1992)
Histoplasma capsulatum	Woods *et al.* (1993)
Lactococcus lactis	Cancilla *et al.* (1992)
Leptosphaeria maculans	Goodwin and Annis (1991)
Listeria monocytogenes	Czajka *et al.* (1993)
Porphyromonas gingivalis	Menard *et al.* (1992)
Rhizobium spp.	Harrison *et al.* (1992)
Staphylococcus spp.	Welsh and McClelland (1990)
Streptococcus spp.	Welsh and McClelland (1990)
Streptococcus uberis	Jayarao *et al.* (1992)

4.3 OTHER AMPLIFICATION METHODS

A number of amplification methods apart from PCR, either for amplifying the target nucleic acid or for amplifying the signal generated in hybridization reactions, have also been developed. While PCR has so far proved to be the method of choice for amplifying target DNA for subsequent identification and/or typing of microorganisms, the alternative methods described in this section offer advantages over PCR in certain areas and are starting to receive increasing attention. Some of these alternative methods are now being developed commercially.

4.3.1. Qβ AMPLIFICATION ASSAYS

This method is named after the Qβ bacteriophage and utilizes the RNA directed RNA polymerase (replicase) specific for this organism (Haruna and Spiegelman, 1965).

Apart from PCR, one possible strategy for overcoming the problems of sensitivity associated with identifying the presence of specific microorganisms by conventional hybridization tests is to link each probe molecule to a replicable RNA molecule that can be amplified exponentially, following hybridization, to reveal the presence of the bound probe (Chu, Kramer and Orgel, 1986). The Qβ replicase enzyme has been incorporated into this strategy by embedding a probe sequence within the sequence of a small, naturally-occurring RNA template (termed MDV–1) for Qβ replicase. MDV–1 is available as a cDNA sequence forming part of a plasmid which can serve as a template for the synthesis of MDV–1 RNA when incubated *in vitro* with T7 RNA polymerase (Lizardi *et al.*, 1988). The insertion of probe sequences within the cDNA MDV–1 sequence thereby creates templates that can be used for amplification by Qβ replicase. Thus, following hybridization bound probe molecules are recognized by transcription with T7 RNA polymerase, followed by specific amplification with Qβ replicase. Various labels for detection purposes can be incorporated directly into the amplified probe molecules during the replication process (Lizardi *et al.*, 1988).

Qβ amplification assays differ from PCR in that the amplification reaction serves to increase the amount of signal (by increasing the number of labelled probe-derived RNA molecules associated with the target), rather than by increasing the number of target molecules themselves. No thermal cycling is necessary for Qβ amplification assays since Qβ replicase can amplify a single MDV–1 RNA molecule to 10^6–10^9 copies in c.15 min at 37°C (Knight, 1989; Cahill, Foster and Mahan, 1991). The kinetics of the reaction allow the construction of a standard curve and quantification of the signal generated (Walker and Dougan, 1989). Further modifications of the method have been described by Kramer *et al.* (1991) and Wolcott (1992). Although identification tests based on amplification reactions with Qβ replicase are not yet available commercially, the method has already been applied successfully to the detection of cytomegalovirus and human immunodeficiency virus (Lawrence, Braese and Decker, 1990).

4.3.2 TRANSCRIPT AMPLIFICATION SYSTEMS

The transcript amplification system (*TAS*), as described by Kwoh *et al.* (1989), allows the amplification of denatured DNA or a strand of RNA. In the basic method, a primer containing a promoter sequence at the 5′-end for RNA polymerase is used in conjunction with reverse transcriptase to generate a cDNA copy of the target RNA. After heat denaturation, a second primer complementing the RNA polymerase promoter sequence is annealed to the new DNA strand. Reverse transcriptase is then added to produce a dsDNA molecule. RNA polymerase can then be used to produce multiple copies (10 – 1000) of RNA. Only four cycles of the reaction are necessary to obtain c. 10^6 copies of the original template. Six cycles of TAS have been shown to be capable of detecting one lymphocyte infected with human immunodeficiency virus in a field of 10^6 uninfected cells (Gingeras *et al.*, 1990a).

A variation of TAS has been termed **self-sustained sequence repli-**

cation (Gingeras *et al.*, 1990b; Guatelli *et al.*, 1990). This procedure, also known as **3SR amplification**, differs from TAS in that the reaction occurs completely at 37–42°C and the RNA–DNA duplexes formed are degraded by the addition of RNAase H. This allows for conversion to a dsDNA with a polymerase-binding site at each end. This molecule then continues the cycle by functioning as a template for further RNA synthesis. A 10^5-fold amplification occurs within 15 min at 37°C. 3SR will also amplify a DNA target if an initial denaturing step is included. 3SR amplification has been used successfully to monitor drug-resistant mutants of human immunodeficiency virus (Gingeras *et al.*, 1991) but, like PCR and other amplification methods, 3SR may suffer from problems caused by contamination with extraneous nucleic acids, and may also be affected by the lower stringencies and multiple enzymes involved in the reaction.

A further commercial development of TAS (Kievits *et al.*, 1991; Malek *et al.*, 1992), termed **nucleic acid sequence-based amplification (NASBA)**, has been reported to be capable of detecting ten human immunodeficiency virus particles or five infected cells (Kievits *et al.*, 1991).

4.3.3 LIGASE-MEDIATED AMPLIFICATION SYSTEM

The **ligase chain reaction (LCR)** is based on the use of a thermostable DNA ligase from *T. aquaticus* to join two adjacent oligonucleotides that hybridize uniquely to one strand of the target DNA. The junction of the two primers is positioned so that the nucleotide at the 3′-end of the upstream primer coincides with a potential difference of a single bp in the target sequence that defines different species or subtypes. As with PCR, the reaction is cyclic, with repeated steps of denaturation, annealing and ligation. In each cycle, the ligated oligonucleotide pairs, along with the original sequence, form templates for the succeeding cycle (Wu and Wallace, 1989). If there is a mismatch at the primer junction, the primers will not be ligated, and the amplification reaction will not proceed.

A major advantage of the system is that it can reliably distinguish single bp mismatches in microorganisms, but this is balanced by the disadvantage that the entire target sequence must be known in advance. However, it has been reported that successful amplification can result in the detection of as few as ten target molecules following 20–30 cycles of the reaction (Backman, 1992). The method has been used to detect papillomavirus (Bond *et al.*, 1990) and to discriminate *Listeria monocytogenes* from other *Listeria* spp. (Wiedmann *et al.*, 1992).

A modification of the ligase reaction involves the use of a ligase in conjunction with a polymerase (Birkenmeyer and Mushahwar, 1991).

This modification has been termed **gapped-LCR**, and involves the annealing of two oligonucleotide primers to the target in positions that leave a gap between them. DNA polymerase fills this gap between the two primers, and the ligase then connects the filled gap and the second primer.

4.3.4 OTHER SIGNAL AMPLIFICATION METHODS

The sensitivity of all hybridization-based methods for the identification and typing of microorganisms can be enhanced by amplifying the signal attached to the probe. At the present time it is not possible to detect a single hybridized and labelled probe molecule, but a number of methods are available that can provide very substantial signal amplification. Apart from the Qβ system (section 4.3.1.), the most important systems include the application of compound probes, enzymic cycling, and clustering of fluorescent or luminescent labels (Matthews and Kricka, 1988; Dahlen *et al.*, 1989; Persing and Landry, 1989; Soini, 1989).

Although some of these methods can achieve extremely high levels of sensitivity, this high sensitivity may of itself cause problems in interpretation, particularly in the area of diagnostic microbiology. Thus, in clinical or veterinary microbiology, potential pathogens may often be present in small numbers as harmless commensals. How many organisms need to be present to cause disease? In such situations it seems that quantitative hybridization methods, correlated with clinical symptoms, may sometimes be more important than achieving ultimate sensitivity.

4.4 NUCLEIC ACID SEQUENCING TECHNIQUES

The most specific and informative method for identifying and typing microorganisms is to determine the precise nucleotide sequence of definitive regions of the chromosome. Sequencing methods have developed so rapidly in recent years that comparative sequencing of homologous genes is now a standard technique in molecular systematics and phylogenetic studies (Ludwig, 1991). Conserved genes have been sequenced to establish the position of particular organisms in an overall taxonomic scheme (e.g. Kurtzman, 1992), while less-conserved genes can be used to distinguish between individual organisms that are closely related.

4.4.1 BASIC PRINCIPLES OF NUCLEIC ACID SEQUENCING

All rapid nucleic acid sequencing techniques are designed to generate a series of nested labelled fragments with one common end and the

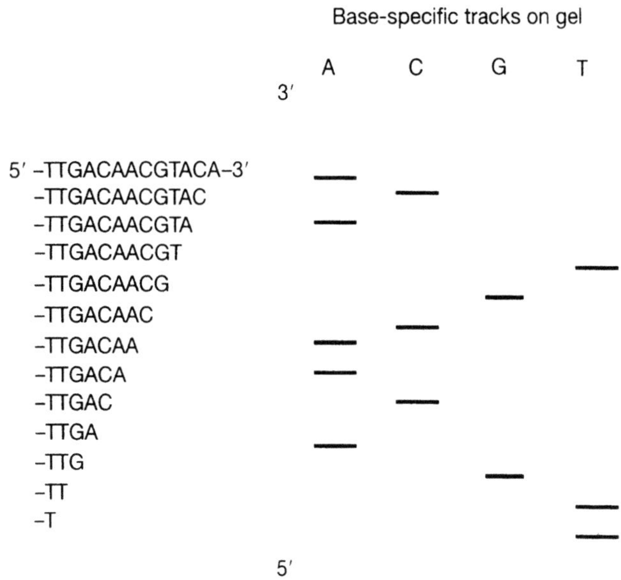

Figure 4.3 Schematic representation of part of a sequencing gel.

other end generated in a base-specific manner. The four base-specific sets of fragments are then separated in parallel by electrophoresis on a high-resolution polyacrylamide slab gel. Such gels can resolve fragments differing in length by a single nucleotide. The four gel lanes (each corresponding to one of the four bases, adenine, cytosine, guanine, or thymine) should therefore contain a series of labelled bands, ideally with only one of the four tracks containing a band at each level. Reading upwards from the bottom of the gel gives the sequence of the DNA strand being examined (Figure 4.3). Several methods are available for generating the sets of base-specific DNA fragments, of which the methods used most widely are the chemical (Maxam and Gilbert, 1977) and chain termination (Sanger *et al.*, 1977, 1978) methods.

(a) Chemical (Maxam and Gilbert) sequencing

This procedure was first described in 1977, and then in more detail by Maxam and Gilbert (1980). It is this latter publication which forms the basis for most protocols used today.

The basis of the chemical sequencing technique is the base-specific or base-selective partial cleavage of end-labelled DNA. The subject DNA is modified and partially cleaved in a base-specific manner, with at least four different reactions to allow differentiation of the four bases. Maxam and Gilbert (1977) described reactions that cleave DNA preferentially at guanines, at cytosines, at adenines and guanines equally, and at thymines and cytosines equally. Many modifications have been made subsequently to optimize the sequencing reactions and improve the resolution of the fragments. A protocol allowing 600 bp sequences to be elucidated from one metre long gel systems has been described in detail by Barker (1989).

The chemical method of DNA sequencing is still preferred by many research workers throughout the world. Its main advantages are that ssDNA can be sequenced directly, sequencing reactions are not affected by long runs of the same nucleotide, and procedures of cloning or subcloning are minimized, thereby avoiding problems of instability that may be encountered with certain vectors used in chain termination sequencing (see below). The main disadvantages of the chemical sequencing method are the complexity of the various chemical reactions and the time required.

(b) Chain termination (Sanger) sequencing

The chain termination method relies on the in-vitro copying of a DNA template by elongation from a site-specific oligonucleotide primer, normally with either Klenow fragment or a T7-derived DNA polymerase, and the four deoxyribonucleotide triphosphates (dNTPs). Four parallel reactions are carried out; in each reaction, one of the dNTPs is substituted partly by its dideoxy analogue, thereby resulting in base-specific chain termination following random incorporation of the dideoxy analogue instead of the correct base. The chain termination method was designed originally for sequencing ssDNA, typically cloned in bacteriophage M13 (Sanger *et al.*, 1978), but protocols for sequencing double-stranded plasmids containing cloned DNA fragments of interest have also been developed (e.g. Chen and Seeburg, 1985; Agellan and Chen, 1986; Williams *et al.*, 1986). Recent modified protocols and sequencing strategies for the chain termination method have been described in detail by Bankier and Barrell (1989).

Despite the advantages of the chemical method for DNA sequencing described in the previous section, so far as sequence determination for microbial identification and typing purposes is concerned, the chain termination method has become established as the method of choice. In part this is because the chain termination method is simpler and less time-consuming than the chemical method. In addition, the genes used for comparative sequence analysis in identification and typing applications usually contain highly conserved and less conserved regions, and therefore primers complementary to highly conserved genes can be used to analyse a range of different organisms.

4.4.2 SEQUENCING OF PCR PRODUCTS

The main problems preventing widespread use of DNA sequencing for identification and typing purposes have been the complexity and time-consuming nature of the procedures, the small amount of target material present in most samples, and the necessity for cloning procedures in the chain termination method. Fortunately, PCR products can now be sequenced directly, thereby avoiding the necessity for cloning procedures, and considerably speeding-up and simplifying the procedure (Gyllensten, 1989).

Several methods are available for direct sequencing of double-stranded PCR products by the chain termination method (Wrischnik *et al.*, 1987; Levedakou, Landegren and Hood, 1989; Winship, 1989), but the most popular way of generating template DNA for sequencing purposes is to use a technique termed '**asymmetric priming**' to synthesize an excess of one strand by PCR (Gyllensten and Erlich, 1988). In this method, one of the primers is limited in concentration so that, in the presence of only one effective primer, the PCR will no longer proceed exponentially, but will produce numerous single-stranded copies from the primer provided in excess.

Two technical problems are associated with direct chain termination sequencing of amplification products. Firstly, excess of primers and dNTPs present during the amplification reaction must be removed to avoid interference with the sequencing reactions. Secondly, re-annealing of the dsDNA fragment will compete with the annealing reaction between the DNA strand to be sequenced and the sequencing primer. Syvänen and Söderlund (1991) have described a solid-phase technique to circumvent these problems in which the sequence of interest is first amplified by PCR with a primer biotinylated at its 5′end. The amplified 5′ biotinylated fragments are then captured on avidin-coated polystyrene microparticles, thereby facilitating purification of the amplified DNA from the PCR components and the non-biotinylated second strand of the amplified fragment. Chain termination reactions can then be performed with the immobilized single-stranded template, followed by automated sequence elucidation.

4.5 SUMMARY

The availability of nucleic acid amplification techniques, at the present time chiefly PCR, is in the process of revolutionizing many procedures in molecular biology, including those used for identifying and typing microorganisms. PCR is now used widely and routinely in specialist laboratories to rapidly identify the presence of particular microorganisms in unpurified samples by amplifying a specific diagnostic segment of the genome selectively. New variations on the basic PCR method are described on an almost daily basis, and it is likely that the alternative amplification methods outlined in this chapter may provide even more advantages and opportunities for creative scientific thought following their commercial development.

So far as typing of microorganisms is concerned, the most precise typing is accomplished by identification of nucleotide sequence variations, and a rapid diagnostic sequencing technique would be extremely useful for epidemiological studies. PCR provides a rapid method for producing relatively large quantities of template material for sequencing from a selected locus in genomic DNA, thereby allowing work which would normally take weeks by conventional procedures to be performed in a few hours. Voss *et al.* (1989) have expanded on the possibilities by describing a technique in which DNA is amplified by PCR in combination with one fluorescence-labelled and one unlabelled primer. The amplified DNA is neither purified nor cloned, but is simply bound to a membrane and (base-specifically) cleaved chemically *in situ*. The sequence is then determined on an automated sequencing system in which the migration order of the fluorescently labelled DNA fragments is detected by a laser coupled to an on-line computer.

Although methods have now been available for determining sequence variations in microorganisms for over 15 years, the complexity of these methods, and the technical skills required, have meant that they have been used mainly in research applications defining the position of microorganisms in an overall taxonomic structure. Recent developments in sequencing technology and amplification methods mean that a rapid sequencing technique for typing purposes is now a realistic goal. The availability of rapid and convenient methods for determining sequence variations between different microorganisms seem likely to make such methods increasingly important tools for identification and typing of microorganisms in the near future.

REFERENCES

Agellon, L.B. and Chen, T.T. (1986) Supercoiled plasmid sequencing: one buffer for template denaturation, primer annealing and synthesis of labelled strands by reverse transcriptase. *Gene Analysis Techniques*, **3**, 86–9.

Akopyanz, N., Bukanov, N.O., Westblom, T.U. *et al.* (1992) DNA diversity among clinical isolates of *Helicobacter pylori* detected by PCR-based RAPD fingerprinting. *Nucleic Acids Research*, **20**, 5137–42.

Altamirano, M., Kelly, M.T., Wong, A. *et al.* (1992) Characterization of a DNA probe for detection of *Mycobacterium tuberculosis* complex in clinical samples by polymerase chain reaction. *Journal of Clinical Microbiology*, **30**, 2173–6.

An. S.F., Franklin, D. and Fleming, K.A. (1992) Generation of digoxigenin-labelled double-stranded and single-stranded probes using the polymerase chain reaction. *Molecular and Cellular Probes*, **6**, 193–200.

Aufauvre-Brown, A., Cohen, J. and Holden, D.W. (1992) Use of randomly amplified polymorphic DNA markers to distinguish isolates of *Aspergillus fumigatus*. *Journal of Clinical Microbiology*, **30**, 2991–3.

Backman, K. (1992) Ligase chain reaction: diagnostic technology for the 1990s and beyond. *Clinical Chemistry*, **38**, 457–8.

Bacot, C.M. and Reeves, R.H. (1991) Novel tRNA gene organization in the 16S–23S intergenic spacer of the *Streptococcus pneumoniae* rRNA gene cluster. *Journal of Bacteriology*, **173**, 4234–6.

Bankier, A.T. and Barrell, B.G. (1989) Sequencing single-stranded DNA using the chain termination method, in *Nucleic Acids Sequencing: A Practical Approach* (eds C.J. Howe and E.S. Ward), IRL Press, Oxford, pp. 37–78.

Barker, R.F. (1989) Maxam and Gilbert sequencing using one metre gel systems, in *Nucleic Acids Sequencing: A Practical Approach* (eds. C.J. Howe and E.S. Ward), IRL Press, Oxford, pp. 117–35.

Barker, R.H., Banchongaksorn, T., Courval, J.M. *et al.* (1992) A simple method to detect *Plasmodium falciparum* directly from blood samples using the polymerase chain reaction. *American Journal of Tropical Medicine and Hygiene*, **46**, 416–26.

Bej, A., Mahbubani, M.H. and Atlas, R.M. (1991) Amplification of nucleic acids by polymerase chain reaction (PCR) and other methods and their applications. *Clinical Reviews in Biochemistry and Molecular Biology*, **26**, 301–34.

Bereswill, S., Pahl, A., Bellemann, P. *et al.* (1992) Sensitive and species-specific detection of *Erwinia amylovora* by polymerase chain reaction analysis. *Applied and Environmental Microbiology*, **58**, 3522–6.

Birkenmeyer, L.G. and Mushahwar, I.K. (1991) DNA probe amplification methods. *Journal of Virological Methods*, **35**, 117–26.

Bond, S., Carrino, J., Hampl, H. *et al.* (1990) New methods of detection of HPV, in *Papillomaviruses in Human Pathology. Recent Progress in Epidermal Precancer*, vol. 78 (ed J. Monsenego), Raven Press, New York, pp. 425–34.

Breniere, S.F., Bosseno, M.F., Revollo, S. *et al.* (1992) Direct identification of *Trypanosoma cruzi* natural clones in vectors and mammalian hosts by polymerase chain reaction amplification. *American Journal of Tropical Medicine and Hygiene*, **46**, 335–41.

Brooks, J.L., Moore, A.S., Patchett, R.A. *et al.* (1992) Use of the polymerase chain reaction and oligonucleotide probes for the rapid detection and identification of *Carnobacterium* species from meat. *Journal of Applied Bacteriology*, **72**, 294–301.

Brousseau, R., Saint-Onge, A., Préfontaine, G. *et al.* (1993) Arbitrary primer polymerase chain reaction, a powerful method to identify *Bacillus thuringiensis* serovars and strains. *Applied and Environmental Microbiology*, **59**, 114–19.

Caetano-Annollés, G., Bassam, B.J. and Gresshoff, P.M. (1991) DNA amplification fingerprinting using very short arbitrary oligonucleotide primers. *BioTechnology*, **9**, 553–7.

Cahill, P., Foster, K. and Mahan, D.E. (1991) Polymerase chain reaction and Qβ replicase amplification. *Journal of Virology*, **37**, 1482–5.

Cancilla, M.R., Powell, I.B., Hillier, A.J. and Davidson, B.E. (1992) Rapid genomic fingerprinting of *Lactococcus lactis* strains by arbitrarily primed polymerase chain reaction with P–32 and fluorescent labels. *Applied and Environmental Microbiology*, **58**, 1772–5.

Charlotte, F., Olivier, J.L., Chypre, C. *et al.* (1991) Detection and typing of human papillomaviruses in cervical smears by an original application of the polymerase chain reaction. *Molecular and Cellular Probes*, **5**, 445–50.

Chen, E.Y. and Seeburg, P.H. (1985) Supercoil sequencing: a fast and simple method for sequencing plasmid DNA. *DNA*, **4**, 165–70.

Chu, B.C.F., Kramer, F.R. and Orgel, L.E. (1986) Synthesis of an amplifiable reporter RNA for bioassays. *Nucleic Acids Research*, **14**, 559–603.

Claas, E.C.J., Sprenger, M.J.K., Kleter, G.E.M. *et al.* (1992) Type-specific identification of influenza virus A, Virus B and virus C by the polymerase chain reaction. *Journal of Virological Methods*, **39**, 1–13.

Clayton, C.L., Kleanthous, H., Coates, P.J. *et al.* (1992) Sensitive detection of *Helicobacter pylori* by using polymerase chain reaction. *Journal of Clinical Microbiology*, **30**, 192–200.

Crowhurst, R.N., Hawthorne, B.T., Rikkerink, E.H.A. and Templeton, M.D. (1991) Differentiation of *Fusarium solani* f. sp. *cucurbitae* races 1 and 2 by random amplification of polymorphic DNA. *Current Genetics*, **20**, 391–6.

Czajka, J., Bsat, N., Piani, M. *et al.* (1993) Differentiation of *Listeria monocytogenes* and *Listeria innocua* by 16S rRNA genes and intraspecies discrimination of *Listeria monocytogenes* strains by random amplified polymorphic DNA polymorphisms. *Applied and Environmental Microbiology*, **59**, 304–8.

Dahlén, P.O., Hurskainen, P.J. and Lövgren, T.N.A. (1989) Alternative labels in DNA hybridization, in *Rapid Methods and Automation in Microbiology and Immunology* (eds. A. Balows, R.C. Tilton, and A. Turano), Brixia Academic Press, Brescia, pp. 213–21.

Dawood, M.R., Allan, R., Fowke, K. *et al.* (1992) Development of oligonucleotide primers and probes against structural and regulatory genes of human immunodeficiency virus type–1 (HIV–1) and their use for amplification of HIV–1 provirus by using polymerase chain reaction. *Journal of Clinical Microbiology*, **30**, 2279–83.

de Bruijn, F.J. (1992) Use of repetitive (repetitive extragenic palindromic and enterobacterial repetitive intergenic consensus) sequences and the polymerase chain reaction to fingerprint the genomes of *Rhizobium meliloti* isolates and other soil bacteria. *Applied and Environmental Microbiology*, **58**, 2180–7.

Engstrand, L., Nguyen, A.M.H., Graham, D.Y. and Elzaatari, F.A.K. (1992) Reverse transcription and polymerase chain reaction amplification of rRNA for detection of *Helicobacter* species. *Journal of Clinical Microbiology*, **30**, 2295–301.

Erlich, H.A. (1989) *PCR Technology – Principles and Applications of DNA amplification*. Stockton Press, New York.

Erlich, H.A., Gelfand, D. and Sninsky, J.J. (1991) Recent advances in the polymerase chain reaction. *Science*, **252**, 1643–51.

Fekete, A., Bantle, J.A., Halling, S.M. and Stich, R.W. (1992) Amplification fragment length polymorphism in *Brucella* strains by use of polymerase chain reaction with arbitrary primers. *Journal of Bacteriology*, **174**, 7778–83.

Foxall, P.A., Hu, L.T. and Mobley, H.L.T. (1992) Use of polymerase chain reaction-amplified *Helicobacter pylori* urease structural genes for differentiation of isolates. *Journal of Clinical Microbiology*, **30**, 739–41.

Garson, J.A., Tedder, R.S., Briggs, M. *et al.* (1990) Detection of hepatitis C viral sequences in blood donations by 'nested' polymerase chain reaction and prediction of infectivity. *Lancet*, **335**, 1419–22.

Gingeras, T.R., Richman, D.D., Kwoh, D.Y., and Guatelli, J.C. (1990a) Methodologies for in vitro nucleic acid amplification and their applications. *Veterinary Microbiology*, **24**, 235–51.

Gingeras, T.R., Whitfield, K.M. and Kwoh, D.Y. (1990b) Unique features of the self-sustained replication (3SR) reaction in the in vitro amplification of nucleic acids. *Annales de Biologie Clinique*, **43**, 498–501.

Gingeras, T.R., Prodanovich, P., Latimer, T. *et al.* (1991) Use of self-sustained sequence replication amplification reaction to analyze and detect mutations in zidovudine-resistant human immunodeficiency virus. *Journal of Infectious Diseases*, **164**, 1066–74.

Giovannoni, S. (1991) The polymerase chain reaction, in *Nucleic Acid Techniques in Bacterial Systematics* (eds E. Stackebrandt and M. Goodfellow), Wiley, Chichester, pp. 177–203.

Goodwin, P.H. and Annis, S.L. (1991) Rapid identification of genetic variation and pathotype of *Leptosphaeria maculans* by random amplified polymorphic DNA assay. *Applied and Environmental Microbiology*, **57**, 1482–6.

Gottlieb, P. and Rudner, R. (1985) Restriction site polymorphism of ribosomal ribonucleic acid gene sets in members of the genus *Bacillus*. *International Journal of Systematic Bacteriology*, **35**, 244–52.

Gozlan, J., Caburet, F., Tancrede, C. and Petit, J.C. (1992) A reverse polymerase chain reaction method for detection of human cytomegalovirus late transcripts in cells infected in vitro. *Journal of Virological Methods*, **40**, 1–10.

Grimprel, E., Sanchez, P.J., Wendel, G.D. *et al.* (1991) Use of polymerase chain reaction and rabbit infectivity testing to detect *Treponema pallidum* in amniotic fluid, fetal and neonatal sera, and cerebrospinal fluid. *Journal of Clinical Microbiology*, **29**, 1711–8.

Guatelli, J.C., Whitfield, K.M., Kwoh, D.Y. *et al.* (1990) Isothermal, in vitro amplification of nucleic acids by a multienzyme reaction modeled after retroviral replication. *Proceedings of the National Academy of Sciences of the USA*, **87**, 1874–8.

Gumerlock, P.H., Tang, Y.J., Meyers, F.J. and Silva, J. (1991) Use of the polymerase chain reaction for the specific and direct detection of *Clostridium difficile* in human feces. *Reviews of Infectious Diseases*, **13**, 1053–60.

Gurtler, V., Wilson, V.A., and Mayall, B.C. (1991) Classification of medically important clostridia using restriction endonuclease site differences of PCR-amplified 16S rDNA. *Journal of General Microbiology*, **137**, 2673–9.

Gyllensten, U. (1989) Direct sequencing of in vitro amplified DNA, in *PCR Technology – Principles and Applications for DNA Amplification* (ed H.A. Erlich), Stockton Press, New York, pp. 45–60.

Gyllensten, U.B. and Erlich, H.A. (1988) Generation of single-stranded DNA by the polymerase chain reaction and its application to direct sequencing of the HLA-DQA gene locus. *Proceedings of the National Academy of Sciences of the USA*, **85**, 7652–6.

Hackel, C., Houard, S., Potaels, F. *et al.* (1990) Specific identification of *Mycobacterium leprae* by the polymerase chain reaction. *Molecular and Cellular Probes*, **4**, 205–10.

Harrison, S.P., Mytton, L.R., Skot, L. *et al.* (1992) Characterisation of *Rhizobium* isolates by amplification of DNA polymorphisms using random primers. *Canadian Journal of Microbiology*, **38**, 1009–15.

Haruna, I. and Spiegelman, S. (1965) Specific template requirements of RNA replicases. *Proceedings of the National Academy of Sciences of the USA*, **54**, 579–87.

Hayden, J.D., Ho, S.A., Hawkey, P.M. *et al.* (1991) The promises and pitfalls of PCR. *Reviews in Medical Microbiology*, **2**, 129–37.

Hayes, L.J., Bailey, R.L., Mabey, D.C.W. *et al.* (1992) Genotyping of *Chlamydia trachomatis* from a trachoma-endemic village in The Gambia by a nested polymerase chain reaction identification of strain variants. *Journal of Infectious Diseases*, **166**, 1173–7.

Henchal, E.A., Polo, S.L., Vorndam, V. *et al.* (1991) Sensitivity and specificity of a universal primer set for the rapid diagnosis of Dengue Virus infections by polymerase chain reaction and nucleic acid hybridization. *American Journal of Tropical Medicine and Hygiene*, **45**, 418–28.

Henson, J.M. (1992) DNA hybridization and polymerase chain reaction (PCR) tests for identification of *Gaeumannomyces*, *Phialophora* and *Magnaporthe* isolates. *Mycological Research*, **96**, 629–36.

Herman, L. and Deridder, H. (1992) Identification of *Brucella* spp. by using the polymerase chain reaction. *Applied and Environmental Microbiology*, **58**, 2099–101.

Hill, W.E., Keasler, S.P., Trucksess, M.W. *et al.* (1991) Polymerase chain reaction identification of *Vibrio vulnificus* in artificially contaminated oysters. *Applied and Environmental Microbiology*, **57**, 707–11.

Houard, S., Hackel, C., Herzog, A. and Bollen, A. (1989) Specific identification of *Bordetella pertussis* by the polymerase chain reaction. *Research in Microbiology*, **140**, 477–87.

Innis, M.A., Gelfand, D.H., Sninsky, J.J. and White, T. (1990) *PCR Protocols – a Guide to Methods and Applications*. Academic Press, San Diego.

Innis, M.A., Myambo, K.B., Gelfand, D.H. and Brow, M.A.D. (1988) DNA sequencing with *Thermus aquaticus* DNA polymerase, and direct sequencing of PCR amplified DNA. *Proceedings of the National Academy of Sciences of the USA*, **85**, 9436–40.

Jackson, M.P. (1992) Identification of shiga-like toxin type-II producing *Escherichia coli* using the polymerase chain reaction and a digoxigenin labelled DNA probe. *Molecular and Cellular Probes*, **6**, 209–14.

Jackson, D.P., Quirke, P., Lewis, F. *et al.* (1989) Detection of measles virus RNA in paraffin-embedded tissue. *Lancet*, **i**, 1391.

Jayarao, B.M., Dore, J.J.E., Baumbach, G.A. *et al.* (1991) Differentiation of *Streptococcus uberis* from *Streptococcus parauberis* by polymerase chain reaction and restriction fragment length polymorphism analysis of 16S ribosomal DNA. *Journal of Clinical Microbiology*, **29**, 2774–8.

Jayarao, B.M., Bassam, B.J., Caetanoanolles, G. *et al.* (1992) Subtyping of *Streptococcus uberis* by DNA identification fingerprinting. *Journal of Clinical Microbiology*, **30**, 1347–50.

Jeffreys, A.J., Wilson, V.T. and Thein, S.L. (1985) Hypervariable minisatellite regions in human DNA. *Nature*, **314**, 67–73.

Jeffreys, A.J., Wilson, V., Neumann, R. and Keyte, J. (1988) Amplification of human minisatellites by the polymerase chain reaction: towards DNA fingerprinting of single cells. *Nucleic Acids Research*, **16**, 10 953–71.

Jinks-Robertson, S. and Nomura, M. (1987) Ribosomes and tRNA, in *Escherichia coli and Salmonella typhimurium: Cellular and Molecular Aspects* (eds F.C. Niedhardt *et al.*), American Society for Microbiology, Washington, pp. 1358–85.

Kaltenboeck, B., Kousoulas, K.G. and Storz, J. (1992) Two-step polymerase chain reactions and restriction endonuclease analyses detect and differentiate *ompA* DNA of *Chlamydia* spp. *Journal of Clinical Microbiology*, **30**, 1098–104.

Khush, R.S., Becker, E. and Wach, M. (1992) DNA amplification polymorphisms of the cultivated mushroom *Agaricus bisporus*. *Applied and Environmental Microbiology*, **58**, 2971–7.

Kievits, T., van Gemen, B., van Strijp, D. *et al.* (1991) NASBA isothermal enzymatic in vitro nucleic acid amplification optimized for the diagnosis of HIV-1 infection. *Journal of Virological Methods*, **35**, 273–86.

Klijn, N., Weerkamp, A.H., and Devos, W.M. (1991) Identification of mesophilic lactic acid bacteria by using polymerase chain reaction-amplified variable regions of 16S rRNA and specific DNA probes. *Applied and Environmental Microbiology*, **57**, 3390–3.

Knight, P. (1989) Amplifying probe assays with Q-beta replicase. *BioTechnology*, **7**, 609–10.

Kostman, J.R., Edlind, T.D., Lipuma, J.J. and Stull, T.L. (1992) Molecular epidemiology of *Pseudomonas cepacia* determined by polymerase chain reaction ribotyping. *Journal of Clinical Microbiology*, **30**, 2084–7.

Kramer, F.R., Tyagi, S., Guerra, C.E. *et al.* (1991) Qβ amplification assays, in *Rapid Methods and Automation in Microbiology and Immunology* (eds A. Vaheri, R.C. Tilton and A. Balows), Springer-Verlag, Berlin, pp. 17–22.

Kurtman, C.P. (1992) rRNA sequence comparisons for assessing phylogenetic relationships among yeasts. *International Journal of Systematic Bacteriology*, **42**, 1–6.

Kwoh, D.Y., Davis, G.R., Whitfield, K.M. *et al.* (1989) Transcription-based amplification system and detection of amplified human immunodeficiency virus type 1 with a bead-based sandwich hybridization format. *Proceedings of the National Academy of Sciences of the USA*, **86**, 1173–7.

Lawrence, N.P., Braese, B.S. and Decker, R.S. (1990) Application of a rapid nucleic acid hybridization method and oligo (dT) magnetic particles to the quantitative detection of viral nucleic acids. *Diagnostic Clinical Testing*, **28**, 40–4.

Laxer, M.A., Timblin, B.K. and Patel, R.J. (1991) DNA sequences for the specific detection of *Cryptosporidium parvum* by the polymerase chain reaction. *American Journal of Tropical Medicine and Hygiene*, **45**, 688–94.

Lehmann, P.F., Lin, D.M. and Lasker, B.A. (1992) Genotypic identification and characterization of species and strains within the genus *Candida* by using random amplified polymorphic DNA. *Journal of Clinical Microbiology*, **30**, 3249–54.

Levedakou, E.N., Landegren, U. and Hood, L.E. (1989) A strategy to study gene polymorphism by direct sequence analysis of cosmid clones and amplified genomic DNA. *BioTechniques*, **7**, 438–42.

Lizardi, P.M., Guerra, C.E., Lomeli, H. *et al.* (1988) Exponential amplification of recombinant-RNA hybridization probes. *BioTechnology*, **6**, 1197–202.

Ludwig, W. (1991) DNA sequencing in bacterial systematics, in *Nucleic Acid*

Techniques in Bacterial Systematics (eds E. Stackebrandt and M. Goodfellow), Wiley, Chichester, pp. 69–94.

Lungu, O., Wright, T.C. and Silverstein, S. (1992) Typing of human papillomaviruses by polymerase chain reaction amplification with L1 consensus primers and RFLP analysis. *Molecular and Cellular Probes*, **6**, 145–52.

Maiden, M.C.J., Bygraves, J.A., McCarvil, J. and Feavers, I.M. (1992) Identification of meningococcal serosubtypes by polymerase chain reaction. *Journal of Clinical Microbiology*, **30**, 2835–41.

Malek, L., Darasch, S., Davey, C. *et al.* (1992) Application of NASBA isothermal nucleic acids amplification method to the diagnosis of HIV–1. *Clinical Chemistry*, **38**, 458.

Marconi, R.T. and Garon, C.F. (1992) Development of polymerase chain reaction primer sets for diagnosis of Lyme disease and for species-specific identification of Lyme disease isolates by 16S rRNA signature nucleotide analysis. *Journal of Clinical Microbiology*, **30**, 2830–4.

Matthews, J.A. and Kricka, L.J. (1988) Analytical strategies for the use of DNA probes. *Analytical Biochemistry*, **169**, 1–25.

Maxam, A.M. and Gilbert, W. (1980) Sequencing end-labeled DNA with base-specific chemical cleavages, in *Methods in Enzymology*, vol. 65 (eds L. Grossman and K. Moldave), Academic Press, London, pp. 499–560.

Mazurier, S., van de Giessen, A., Heuvelman, K. and Wernars, K. (1992) RAPD analysis of *Campylobacter* isolates: DNA fingerprinting without the need to purify DNA. *Letters in Applied Microbiology*, **14**, 260–2.

McClelland, M., Petersen, C. and Welsh, J. (1992) Length polymorphisms in transfer RNA intergenic spacers detected by using the polymerase chain reaction can distinguish streptococcal strains and species. *Journal of Clinical Microbiology*, **30**, 1499–504.

McMillin, D.E. and Muldrow, L.L. (1992) Typing of toxic strains of *Clostridium difficile* using DNA fingerprints generated with arbitrary polymerase chain reaction primers. *FEMS Microbiology Letters*, **92**, 5–10.

McOmish, F., Yap, P.L., Jordan, A. *et al.* (1993) Detection of parvovirus B19 in donated blood: a model system for screening by polymerase chain reaction. *Journal of Clinical Microbiology*, **31**, 323–8.

Menard, C., Brousseau, R. and Mouton, C. (1992) Application of polymerase chain reaction with arbitrary primer (AP-PCR) to strain identification of *Porphyromonas* (*Bacteroides*) *gingivalis*. *FEMS Microbiology Letters*, **95**, 163–8.

Merien, F., Amouriaux, P., Perolat, P. *et al.* (1992) Polymerase chain reaction for detection of *Leptospira* spp. in clinical samples. *Journal of Clinical Microbiology*, **30**, 2219–24.

Miyakawa, Y., Mabuchi, T., Kagaya, K. and Fukazawa, Y. (1992) Isolation and characterization of a species-specific DNA fragment for detection of *Candida albicans* by polymerase chain reaction. *Journal of Clinical Microbiology*, **30**, 894–900.

Mullis, K.B. and Faloona, F.A. (1987) Specific synthesis of DNA in vitro via a polymerase-catalyzed chain reaction, in *Methods in Enzymology*, vol. 155 (ed R. Wu), Academic Press, San Diego, pp. 335–50.

Murakami, K., Minamide, W., Wada, K. *et al.* (1991) Identification of methicillin-resistant strains of staphylococci by polymerase chain reaction. *Journal of Clinical Microbiology*, **29**, 2240–44.

Murakawa, G.J., Zaia, J.A., Spallone, P.A. *et al.* (1988) Direct detection of HIV-1 RNA from AIDS and ARC patient samples. *DNA*, **7**, 287–95.

Nakajima, H., Inoue, M., Mori, T. *et al.* (1992) Detection and identification of *Yersinia pseudotuberculosis* and pathogenic *Yersinia enterocolitica* by an improved polymerase chain reaction method. *Journal of Clinical Microbiology*, **30**, 2484–6.

Navarro, E., Simonet, P., Normand, P. and Bardin, R. (1992) Characterization of natural populations of *Nitrobacter* spp. using PCR/RFLP analysis of the ribosomal intergenic spacer. *Archives of Microbiology*, **157**, 107–15.

Okamoto, H., Sugiyama, Y., Okada, S. *et al.* (1992) Typing hepatitis C virus by polymerase chain reaction with type-specific primers – application to clinical surveys and tracing infectious sources. *Journal of General Virology*, **73**, 673–9.

Olsson, M., Elvin, K., Löfdahl, S. and Linder, E. (1993) Detection of *Pneumocystis carinii* DNA in sputum and bronchoalveolar lavage samples by polymerase chain reaction. *Journal of Clinical Microbiology*, **31**, 221–6.

Ou, C.-Y., Kwok, S., Mitchell, S.W. *et al.* (1988) DNA amplification for direct detection of HIV-1 in DNA of peripheral blood mononuclear cells. *Science*, **239**, 295–7.

Persing, D.H. (1991) Polymerase chain reactions: trenches to benches. *Journal of Clinical Microbiology*, **29**, 1281–5.

Persing, D.H. and Landry, M.L. (1989) In vitro amplification techniques for the detection of nucleic acids: new tools for the diagnostic laboratory. *Yale Journal of Biology and Medicine*, **62**, 159–71.

Porter-Jordan, K., Roseberg, E.I., Keiser, J.F. *et al.* (1990) Nested polymerase chain reaction assay for the detection of cytomegalovirus overcomes false positives caused by contamination with fragmented DNA. *Journal of Medical Virology*, **30**, 85–91.

Rahn, K., Degrandis, S.A., Clarke, R.C. *et al.* (1992) Amplification in an *invA* gene sequence of *Salmonella typhimurium* by the polymerase chain reaction as a specific method of detection of Salmonella. *Molecular and Cellular Probes*, **6**, 271–9.

Regnery, R.L., Spruill, C.L. and Plikaytis, B.D. (1991) Genotypic identification of rickettsiae and estimation of intraspecies sequence divergence for portions of two rickettsial genes. *Journal of Bacteriology*, **173**, 1576–89.

Riley, D.E., Roberts, M.C., Takayama, T. and Krieger, J.N. (1992) Development of a polymerase chain reaction-based diagnosis of *Trichomonas vaginalis*. *Journal of Clinical Microbiology*, **30**, 465–72.

Robertson, N.L., French, R. and Gray, S.M. (1991) Use of group-specific primers and the polymerase chain reaction for the detection and identification of luteoviruses. *Journal of General Virology*, **72**, 1473–7.

Ross, B.C. and Dwyer, B. (1993) Rapid, simple method for typing isolates of *Mycobacterium tuberculosis* by using the polymerase chain reaction. *Journal of Clinical Microbiology*, **31**, 329–34.

Rychlik, W. and Rhoads, R.E. (1989) A computer program for choosing optimal oligonucleotides for filter hybridization, sequencing and in vitro amplification of DNA. *Nucleic Acids Research*, **17**, 8543–51.

Saiki, R.K., Scharf, S., Faloona, F. *et al.* (1985) Enzymatic amplification of beta-globin genomic sequences and restriction site analysis for diagnosis of sickle cell anemia. *Science*, **230**, 1530–4.

Saiki, R.K., Gelfand, D.H., Stoffel, S. *et al.* (1988) Primer-directed enzymatic amplification of DNA with a thermostable DNA polymerase. *Science*, **239**, 487–91.

Samoszuk, M. (1991) Rapid detection of Epstein-Barr viral DNA by nonisotopic in situ hybridization – correlation with the polymerase chain reaction. *American Journal of Clinical Pathology*, **96**, 448–53.

Sanger, F., Nicklen, S. and Coulson, A.R. (1977) DNA sequencing with chain terminating inhibitors. *Proceedings of the National Academy of Sciences of the USA*, **74**, 5463–7.

Sanger, F., Coulson, A.R., Friedmann, T. *et al*. (1978) The nucleotide sequence of bacteriophage ØX174. *Journal of Molecular Biology*, **125**, 225–46.

Schowalter, D.B. and Sommer, S.S. (1989) The generation of radiolabeled DNA and RNA probes with polymerase chain reaction. *Analytical Biochemistry*, **177**, 90–4.

Seal, S.E., Jackson, L.A. and Daniels, M.J. (1992a) Isolation of a *Pseudomonas solanacearum*-specific DNA probe by subtraction hybridization and construction of species-specific oligonucleotide primers for sensitive detection by the polymerase chain reaction. *Applied and Environmental Microbiology*, **58**, 3751–8.

Seal, S.E., Jackson, L.A. and Daniels, M.J. (1992b) Use of tRNA consensus primers to indicate subgroups of *Pseudomonas solanacearum* by polymerase chain reaction amplification. *Applied and Environmental Microbiology*, **58**, 3759–61.

Sellstedt, A., Wullings, B., Nystrom, U. and Gustafsson, P. (1992) Identification of *Casuarina-Frankia* strains by use of polymerase chain reaction (PCR) with arbitrary primers. *FEMS Microbiology Letters*, **93**, 1–5.

Simonet, P., Normand, P., Moiroud, A. and Bardin, R. (1990) Identification of *Frankia* strains in nodules by hybridization of polymerase chain reaction products with strain-specific oligonucleotide probes. *Archives of Microbiology*, **153**, 235–40.

Soini, E.J. (1989) Theoretical comparison of alternative labelling techniques, in *Rapid Methods and Automation in Microbiology and Immunology* (eds A. Balows, R.C. Tilton and A. Turano), Brexia Academic Press, Brescia, pp. 204–12.

Sokol, D.M., Demmler, G.J. and Buffone, G.J. (1992) Rapid epidemiologic analysis of Cytomegalovirus by using polymerase chain reaction amplification of the L-S junction region. *Journal of Clinical Microbiology*, **30**, 839–44.

Srivastava, A.K. and Schlessinger, D. (1990) Mechanism and regulation of ribosomal RNA processing. *Annual Reviews of Microbiology*, **44**, 105–29.

Stein, A. and Raoult, D. (1992) Detection of *Coxiella burnetii* by DNA amplification using polymerase chain reaction. *Journal of Clinical Microbiology*, **30**, 2462–6.

Syvänen, A.C. and Söderlund, H. (1991) Nucleotide sequence determination as a diagnostic tool, in *Rapid Methods and Automation in Microbiology and Immunology* (eds A. Vaheri, R.C. Tilton and A. Balows), Springer-Verlag, Berlin, pp. 23–6.

Taniguchi, K., Wakasugi, F., Pongsuwanna, Y. *et al*. (1992) Identification of human and bovine rotavirus serotypes by polymerase chain reaction. *Epidemiology and Infection*, **109**, 303–12.

Tindall, K.R. and Kunkel, T.A. (1988) Fidelity of DNA synthesis by the *Thermus aquaticus* DNA polymerase. *Biochemistry*, **27**, 6008–13.

van Belkum, A., de Jonckheere, J. and Quint, W.G.V. (1992) Genotyping *Naeg-*

leria spp. and *Naegleria fowleri* isolates by interrepeat polymerase chain reaction. *Journal of Clinical Microbiology*, **30**, 2595–8.

van de Ven, E., Melchers, W., Galama, J. *et al.* (1991) Identification of *Toxoplasma gondii* infections by BI gene amplification. *Journal of Clinical Microbiology*, **29**, 2120–4.

van Eechoutte, M., Rossau, R., Devos, P. *et al.* (1992) Rapid identification of bacteria of the Comamonadaceae with amplified ribosomal DNA restriction analysis (ARDRA). *FEMS Microbiology Letters*, **93**, 227–34.

van Kuppeveld, F.J.M., van de Logt, J.T.M. *et al.* (1992) Genus-specific and species-specific identification of mycoplasmas by 16S ribosomal RNA amplification. *Applied and Environmental Microbiology*, **58**, 2606–15.

Versalovic, J., Koeuth, T. and Lupski, J.R. (1991) Distribution of repetitive DNA sequences in eubacteria and application to fingerprinting of bacterial genomes. *Nucleic Acids Research*, **19**, 6823–31.

Versalovic, J., Koeuth, T., Zhang, Y.-H. *et al.* (1992) Quality control for bacterial inhibition assays: DNA fingerprinting of microorganisms. *Screening*, **1**, 175–83.

Vodkin, M.H., Howe, D.K., Visvesvara, G.S. and McLaughlin, G.L. (1992) identification of *Acanthamoeba* at the generic and specific levels using the polymerase chain reaction. *Journal of Protozoology*, **39**, 378–85.

Vold, B. (1985) Structure and organization of genes for transfer ribonucleic acid in *Bacillus subtilis*. *Microbiology Reviews*, **49**, 71–80.

Voss, H., Schwager, C., Wirkner, U. *et al.* (1989) Direct genomic fluorescent on-line sequencing using in vitro amplification of DNA. *Nucleic Acids Research*, **17**, 2517–27.

Walker, J. and Dougan, G. (1989) DNA probes: a new role in diagnostic microbiology. *Journal of Applied Bacteriology*, **67**, 229–38.

Welsh, J. and McClelland, M. (1990) Fingerprinting genomes using PCR with arbitrary primers. *Nucleic Acids Research*, **18**, 7213–8.

Welsh, J. and McClelland, M. (1991) Species-specific genomic fingerprints produced by PCR with consensus tRNA gene primers. *Nucleic Acid Research*, **19**, 861–6.

Welsh, J. and McClelland, M. (1992) PCR-amplified length polymorphisms in transfer RNA intergenic spacers for categorizing staphylococci. *Molecular Microbiology*, **6**, 1673–80.

Wiedmann, M., Czajka, J., Barany, F. and Batt, C.A. (1992) Discrimination of *Listeria monocytogenes* from other *Listeria* species by ligase chain reaction. *Applied and Environmental Microbiology*, **58**, 3443–7.

Williams, S.A., Slatko, B.E., Moran, L.S. and DeSimone, S.M. (1986) Sequencing in the fast lane: a rapid protocol for (α-35S)dATP dideoxy DNA sequencing. *BioTechniques*, **4**, 138–47.

Williams, J.G.K., Kubelik, A.R., Livak, K.J. *et al.* (1990) DNA polymorphisms amplified by arbitrary primers are useful as genetic markers. *Nucleic Acids Research*, **18**, 6531–5.

Wilson, K.H., Blitchington, R., Shah, P. *et al.* (1989) Probe directed at a segment of *Rickettsia rickettsii* rRNA amplified with polymerase chain reaction. *Journal of Clinical Microbiology*, **27**, 2692–6.

Winship, P.R. (1989) An improved method for directly sequencing PCR amplified material using dimethyl sulfoxide. *Nucleic Acids Research*, **17**, 1266.

Wolcott, M.J. (1992) Advances in nucleic acid-based detection methods. *Clinical Microbiology Reviews*, **5**, 370–86.

Woods, C.R., Versalovic, J., Koeuth, T. and Lupski, J.R. (1992) Analysis of relationships among isolates of *Citrobacter diversus* by using DNA fingerprints generated by repetitive sequence-based primers in the polymerase chain reaction. *Journal of Clinical Microbiology*, **30**, 2921–9.

Woods, J.P., Kersulyte, D., Goldman, W.E. and Berg, D.E. (1993) Fast DNA isolation from *Histoplasma capsulatum*: methodology for arbitrary primer polymerase chain reaction-based epidemiological and clinical studies. *Journal of Clinical Microbiology*, **31**, 463–4.

Wrischnik, L.A., Higuchi, R.G., Stoneking, M. *et al.* (1987) Length mutations in human mitochondrial DNA – direct sequencing of enzymatically amplified DNA. *Nucleic Acids Research*, **15**, 529–42.

Wu, D.Y. and Wallace, R.B. (1989) The ligation amplification reaction (LAR) – amplification of specific DNA sequences using sequential rounds of template-dependent ligation. *Genomics*, **4**, 560–9.

Yang, C.F., De, L., Holloway, B.P. *et al.* (1991) Detection and identification of vaccine-related polioviruses by the polymerase chain reaction. *Virus Research*, **20**, 159–79.

Analysis of protein and lipopolysaccharide profiles

5

5.1. INTRODUCTION TO PROTEIN FINGERPRINTING

As outlined in Chapter 1, typing schemes based on microbial proteins use electrophoresis to generate a fingerprint or profile which allows comparison and differentiation of strains of the organisms of interest. Although the majority of the techniques used do not provide the same degree of discrimination offered by many of the newer nucleic acid based techniques, their widespread use has contributed significantly to the development of typing schemes for a large number of microorganisms.

Proteins used for typing perform a diverse range of structural and metabolic functions in microorganisms. This functional diversity is paralleled by significant variations in the physical properties of the proteins themselves. Protein structure and function is determined by the primary amino acid sequence. Variations in amino acid composition influence secondary and tertiary protein structure and account for differences in the physicochemical properties of protein molecules, including shape, charge, hydrophobicity, isoelectric point, antigenicity and sensitivity or resistance to proteases. In addition, some proteins are bonded covalently to carbohydrates (glycoproteins) of fatty acids (lipoproteins), and these additions also contribute to the overall properties of the molecules. Finally, proteins vary widely in molecular mass, and some, e.g. many enzymes, have a quaternary structure and consist of complexes of several identical or distinct subunits producing molecules with a molecular mass of several hundred thousand daltons.

Many of these physicochemical properties, either singly or in combination, are exploited in the electrophoretic analysis of proteins, and the complexity of the profiles generated varies according to the type of sample used and the analytical method employed.

Whole-cell lysates of microorganisms may contain many hundreds of proteins and, consequently, profiles generated from whole cells are usually complex and may require computer-assisted recording and analysis. Frequently therefore, typing schemes rely on the presence of a relatively small number of major proteins within the organism to allow differentiation by eye. Profiles may be simplified further by using sub-cellular fractions such as membrane extracts, or specific groups of proteins, e.g. bacterial penicillin-binding proteins, for the analysis.

As with nucleic acids, protein molecules present in these samples can be analysed in an intact or native form, e.g. some enzymes, or after physical or enzymic degradation to their component polypeptide or peptide subunits.

This chapter provides an overview of the principles and techniques involved in the generation of protein profiles from microorganisms, and the use of these techniques in typing and identification.

5.2 PROTEIN ANALYSIS

5.2.1. PREPARATION OF SAMPLES FOR PROTEIN ANALYSIS

(a) Growth of microorganisms for analysis

In contrast to DNA composition, the protein content of a microbial cell may be influenced by environmental factors. As an extreme example, inducible enzymes may only be produced in the presence of an appropriate inducing substrate. In addition, for many pathogenic bacteria it is now recognized that expression of virulence-associated determinants is influenced markedly by culture conditions. For example, changes in protein expression associated with variable growth conditions were reported recently for *Pasteurella haemolytica* (Davies *et al.*, 1992). To allow generation of reproducible comparative protein profiles, it is therefore extremely important that growth conditions are standardized. Attention should be paid to inoculum preparation, growth medium formulation, incubation conditions, e.g. time, temperature, gaseous atmosphere, shaken versus static incubation, and the conditions used for harvesting and washing of microbial cells. For some applications, microbial proteins may be radiolabelled with ^{35}S, ^{3}H or ^{14}C by incorporation of a suitable labelled substrate in the growth medium (Tabaqchali *et al.*, 1987; Walpita *et al.*, 1992). Use of radiolabelling introduces an additional safety aspect to these techniques which should be considered when planning the studies.

Phenotypic changes may also occur in some organisms following repeated subculture in the laboratory. For example, the bacterium *Borrelia burgdorferi* loses plasmids which encode outer surface proteins after multiple subculture (Schwan *et al.*, 1988). Similarly, changes in isoenzyme profiles have been reported following subculture of *Candida albicans* (Lehmann *et al.*, 1991). Where possible therefore, stock cultures of organisms for use in studies should ideally be stored frozen and subcultured only when required for analysis.

The quantity of organisms required for different types of analysis varies. For many bacteria, pellets from 1–1.5 ml of an overnight broth culture are usually sufficient for analysis of whole cell profiles. Larger volumes are often required if subcellular fractions, e.g. outer membrane preparations, are to be analysed. Once harvested, microbial cells can be stored frozen at –70°C, enabling large numbers of strains collected over a period of time to be examined simultaneously, thereby increasing the reproducibility of the analysis.

(b) Preparation of protein extracts – solubilization

Irrespective of the analytical procedures used, proteins must be in a soluble state to allow their separation during electrophoresis. Solubilization of proteins involves physical and/or chemical treatment of the microbial cells resuspended in an appropriate volume of a suitable buffer, followed by a centrifugation step to remove remaining whole cells or non-proteinaceous debris which could interfere with protein resolution.

Many cytoplasmic and periplasmic microbial proteins are water-soluble, and these can be obtained readily for analysis by a variety of physical cell disruption procedures such as sonication, or by using a French press or Braun homogenizer (Coakley *et al.*, 1977). To preserve protein structure during extraction, suspensions of microorganisms are chilled on ice during cell breakage. To protect released proteins from degradation by proteases, a protease inhibitor such as phenylmethylsulphonyl fluoride (PMSF) can be added to the buffer used to resuspend the organisms before cell breakage. Use of these physical methods to lyse microorganisms invariably generates aerosols, and suitable safety precautions, including the use of enclosed sonicators and appropriate safety cabinets, is essential, particularly when dealing with large quantities of organisms or pathogens.

Although these physical methods release soluble proteins, they are generally ineffective at solubilizing structural and membrane-associated proteins, which usually requires a combination of chemical and physical treatments. Detergents are used commonly to solubilize lipid-containing membranes into which proteins may be anchored. Non-ionic detergents, such as Triton-X–100 or Nonidet P40, are used where it is necessary to maintain the native charge of the protein during the analysis, e.g. during isoelectric focusing. Anionic detergents, such as sodium dodecyl sulphate (SDS), are used when this is not an important consideration, and are extremely efficient at solubilizing many membrane proteins. Reducing agents, such as 2-mercaptoethanol or dithiothreitol, are usually included in solubilization buffers to reduce disulphide bonds which contribute to the secondary and tertiary structure of protein molecules. Chaotropic agents, such as 5 M urea, are also used commonly in the solubilization of proteins for isoelectric focusing, and disrupt intramolecular forces, such as hydrogen bonds, within the proteins. Finally, the efficiency of solubilization is increased by heating. Boiling microbial cells in the presence of detergents and reducing agents solubilizes the majority of proteins, and simultaneously neutralizes proteases which may be present in the extracts. This procedure usually degrades proteins to their composite polypeptides, and is therefore not applicable to analysis of proteins in their native state. Examples of solubilization protocols for microbial proteins are listed in Table 5.1.

Once proteins have been solubilized, a variety of electrophoresis techniques are available for the analysis of microbial extracts.

Table 5.1 Examples of protein solubilization methods used in protein fingerprinting

Technique	Solubilization method	Reference
MEE	Sonication on ice in 10 mM Tris, 1 mM EDTA, 0.5 mM NaDP (pH 6.8)	Selander *et al.* (1986)
PAGE	Sonication on ice in 50 mM Tris/HCl (pH 7.8)	Matsuda *et al.* (1992)
SDS-PAGE	Boiling for 10 min in 62.5 mM Tris/HCl (pH 6.8), 2% SDS, 5% 2-mercaptoethanol, 10% sucrose and 0.002% bromophenol blue	Hames (1990)
Isoelectric focusing	5 min incubation on ice in 10 mM Tris/HCl (pH 7.4), 1 mM EDTA, 8 M Urea, 50 mM dithiothreitol, 2 mM PMSF, 10% glycerol, 5% Nonidet P40, RNAase A 200 μg/ml and 6% pH 3–5 carrier ampholytes	Cash (1991)

5.2.2. TECHNIQUES USED IN THE GENERATION OF PROTEIN FINGERPRINTS

(a) Multilocus enzyme electrophoresis

Multilocus enzyme electrophoresis (MEE) was developed originally for the analysis of genetic diversity in mammalian populations, and is now used widely for the discrimination of closely-related microorganisms. MEE evolved from earlier techniques of enzyme analysis, usually referred to in the literature as isozyme or isoenzyme electrophoresis. Applications of these original methods to typing and identification of bacteria are reviewed by Williams and Shah (1980). Although MEE and isoenzyme profiling differ in some details (see below), the kind of information they provide for basic typing applications is essentially identical, and nowadays these terms are often used interchangeably. The remainder of this discussion will concentrate on the technique of MEE, which has proven particularly useful in differentiating and typing bacteria for epidemiological purposes. More recently, MEE has also been applied to the analysis of genetic diversity in microbial populations (Selander *et al.*, 1986; Selander and Musser, 1990).

Using MEE, microorganisms are differentiated by analysis of the electrophoretic mobility of a range of soluble basic metabolic enzymes extracted from the organisms. Within a microbial population, several variant forms of individual enzymes (alloenzymes) exist, each encoded by different alleles at a particular genetic locus (isoenzymes, in contrast, may be produced either by different alleles at the same gene locus or from different gene loci). Single amino acid changes in the enzymic proteins encoded by different alleles (or genes) are sufficient to affect their electrophoretic mobility and allow their differentiation. Although information of value for strain typing may be gained by comparing the mobilities of single, or small numbers of enzymes (e.g. Gharbia and Shah, 1990; Mariani-Kurkdjian *et al.*, 1993), this approach may not provide adequate strain differentiation for some applications. The discrimination of the technique can, however, be enhanced markedly by analysis of multiple enzymes, and a panel of 10–15 enzymes is used commonly for MEE. Combined data for the range of enzymes tested is then used to assign an **electrophoretotype (ET)** to a particular isolate or strain, which identifies that organism and allows its comparison with others. Since the enzymes analysed reflect a representative sample of the structural genes of the organism, a combination of multilocus genotypes and allele frequencies can be used to investigate genetic variation within populations, and levels of genetic relatedness between species or isolates (Selander *et al.*, 1986).

As with all other forms of protein electrophoresis, MEE involves sample preparation, electrophoresis and visualization of the resulting electrophoretic profile with appropriate reagents. Detailed descriptions of the methods and reagents used are given by Selander *et al.* (1986) and an overview only will be given here.

Although applied most widely to bacteria, MEE is applicable to any microorganism where sufficient cell biomass can be obtained. Since MEE uses a phenotypic character, care should again be taken to standardize growth conditions for the organisms being analysed. The technique relies on the expression of constitutive enzymes, which ideally should be produced by all isolates of the species under study. The activity and amount of enzyme detected, but not its electrophoretic mobility, may be affected by cultural conditions.

Since MEE relies on the enzymic activity of proteins to allow their visualization, sample preparation uses techniques that conserve the structure and function of the protein molecules. Most commonly, organisms are suspended in ice-cold lysis buffer, the composition of which varies for different organisms, and are broken by sonication. The sonicate is clarified by centrifugation, and stored frozen in aliquots at −70°C before analysis. Storage at low temperature preserves activity of the majority of enzymes for many months, but stability of these preparations may vary between different organisms. Enzyme activity may also be diminished by repeated freezing and thawing of the protein extracts, which should be avoided if possible.

The gel matrix used in MEE is starch. The major advantage of starch over other matrices such as polyacrylamide is that, after electrophoresis, starch gels can be sliced horizontally to provide three or more replicates of the separated proteins. Each replicate can then be used to analyse the activity of a different enzyme, thereby reducing the number of gel runs required to generate electrophoretotypes based on a large number of different enzymes.

Gel and electrophoresis buffers vary considerably in composition for different enzymes, and resolution and enzyme activity may be affected for different enzymes by buffer pH, the type of buffer and, to some extent, the starch concentration. Optimal conditions for individual enzymes are therefore determined on an empirical basis. Similarly, in choosing which enzymes to use for the analysis, some experimentation may be required. Many enzymes may not resolve sufficiently well to allow differentiation of electrophoretic variants, and selection of those that are useful requires preliminary studies.

In contrast to many other protein electrophoresis systems, e.g. SDS-PAGE, samples for MEE are not applied directly into wells cast in the gel matrix. Protein lysates are first applied to small strips of filter paper, which are then inserted vertically into a slit cut in the precast starch gel, which is run horizontally. Up to 20 samples may be compared on a single gel. To improve resolution, gels are cooled and, during electrophoresis, the soluble proteins migrate on the basis of both charge and size. Variant proteins may migrate towards either the anode or cathode.

Following electrophoresis, replicate samples cut from each gel are equilibrated to the appropriate buffer, and substrate added to allow visualization of enzyme activity. The stains used for particular enzymes may be applied as a solution, or in the form of an agar overlay. Gels are incubated at 37°C in the dark, and detection times vary from minutes to several hours for different enzymes.

A schematic representation of the kind of electrophoretic profile generated using MEE is shown in Figure 5.1. Mobilities of enzyme electromorphs from different strains are compared visually on the same gel slice. For MEE it is insufficient to use relative mobility (rf) values to define electromorphs, even when enzymes are separated on the same gel. For each enzyme, distinct electromorphs are numbered in order of decreasing anodal migration (Table 5.2). By combining data for each electromorph over the range of enzymes tested, each isolate is assigned a distinctive numerical electrophoretotype (ET). Where relatively few strains and enzymes are involved, ETs can be compared manually (e.g. Kappos *et al.*, 1992) and strains with similar ETs grouped together. Where more detailed analysis of genetic relatedness of isolates based on ETs is required, numerical taxonomic methods including cluster analysis can be applied (Selander *et al.*, 1986).

Table 5.2 Electrophoretotypes (ETs) generated from enzyme profiles shown in Figure 5.1

	Electromorph for strains 1–10									
Enzyme	1	2	3	4	5	6	7	8	9	10
A	3	2	3	1	3	2	4	1	2	3
B	4	7	6	1	3	4	5	2	5	1
C	3	5	5	4	6	5	4	1	2	2
etc.										

Over the past decade, MEE has been used increasingly as a typing method for medically important bacteria and, in conjunction with other typing schemes, has proven particularly useful in epidemiological studies to mark strains circulating in healthy and infected populations (Selander and Musser, 1990). ETs effectively mark bacterial clones – organisms originally derived from a single parent cell – and MEE has been used extensively in investigations of bacterial population genetics and studies of the geographical distribution and origins of specific

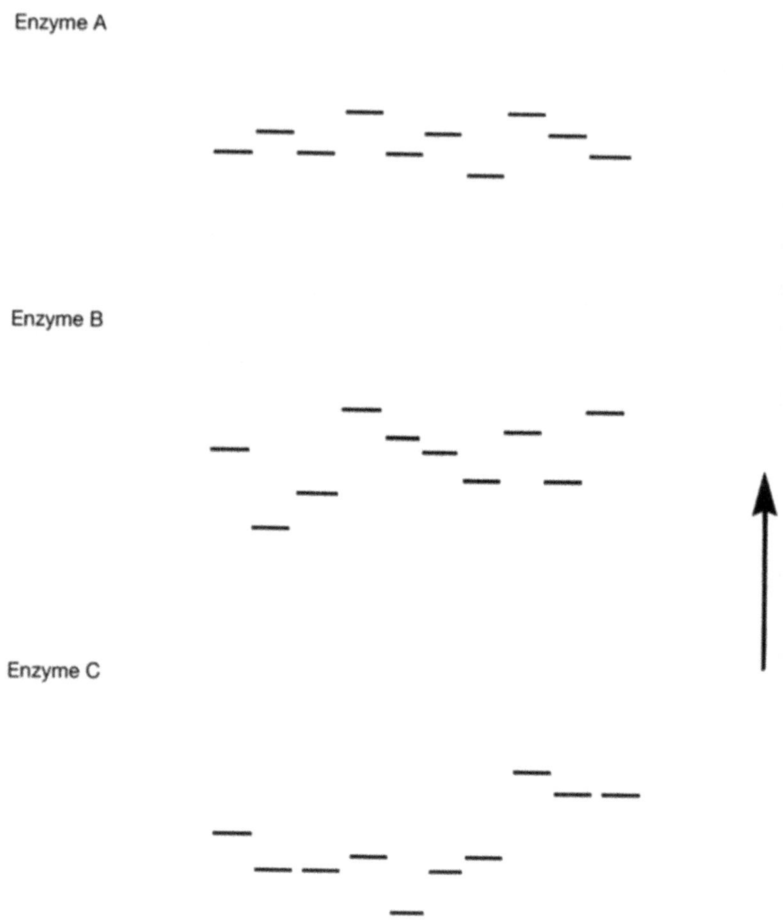

Figure 5.1 Schematic representation of enzyme profiles generated by MEE. Electrophoretograms produced with three enzymes (A), (B) and (C), for ten strains of an organism are shown. The arrow shows the direction of anodal migration from the origin. For each enzyme, each electromorph is numbered in decreasing order of anodal migration to produce the numerical profile shown in Table 5.2. An overall electrophoretype (ET) is assigned by combining data for each strain over the range of enzymes tested.

bacterial pathogens. In cases where epidemics of disease arise, comparison of clones causing the outbreak with others of the same species allows the appearance of new, genetically distinct clones to be recognized. Recently, this approach has also been used in the investigation of the relatedness of bacteria expressing antibiotic resistance in different areas of the world (e.g. Sibold *et al.*, 1992). In addition, typing of bacteria by MEE has allowed associations between expression of known bacterial virulence determinants, e.g. *Haemophilus influenzae* polysaccharide capsules, and disease in particular locations to be assessed (Musser *et al.*, 1988). This type of analysis has also allowed associations between particular bacterial clones and host species to be detected, and has been used to investigate the relatedness of bacteria in disease and carriage states (reviewed by Selander and Musser, 1990).

In addition to its use for epidemiological typing, MEE has also been used for taxonomic studies of bacteria. This technique has proven particularly useful for differentiating genospecies where other methods of phenotypic or serological analysis have lacked reproducibility or been ineffective. For example, MEE detected differences among serotypes of *Legionella pneumophila*, sufficient to warrant subdivision of the species (Selander *et al.*, 1985). Recently, MEE has been used to confirm the existence of distinct genetic types within the single species *B. burgdorferi*, as had been suggested by earlier phenotypic and nucleic acid studies (Boerlin *et al.*, 1992).

Other examples of the use of MEE and other enzyme electrophoresis techniques for typing and epidemiological studies are listed in Table 5.3.

(b) Polyacrylamide gel electrophoresis

Polyacrylamide gel electrophoresis (PAGE) in its various forms is the technique used most often to generate protein, polypeptide or peptide profiles from microorganisms. A detailed description of theoretical and practical applications of PAGE, including the commercial sources of suitable equipment and reagents, is given by Hames (1990), and only an overview of this technique will be presented here.

The gel matrix used – polyacrylamide – consists of a mixture of acrylamide and a cross linking molecule called *bis*-acrylamide. The monomer solution is polymerized by the addition of catalyst, and the resolving properties of the resulting gel are determined by both the total concentration of acrylamide and the relative amount of *bis*-acrylamide present. Conditions for polymerization should be carefully standardized, since variation in the rate of polymerization may affect protein resolution and gel reproducibility. Single concentration gels, ranging from 5–30% acrylamide, can be prepared that allow resolution of molecules ranging in size from hundreds of thousands to only a few thousand daltons. Gradient gels incorporating a linear or concave polyacrylamide gradient, e.g. 5–20%, may also be used, resulting in increased resolution and band sharpening for some molecules, and allowing a wider size range of molecules to be resolved on the same gel. Although

Table 5.3 Examples of organisms for which enzyme electrophoresis has been used for strain differentiation

Organism	Reference
Aeromonas spp.	Tonolla *et al.* (1991)
Aspergillus spp.	Matsuda *et al.* (1992)
Bordetella spp.	Musser *et al.* (1986)
Borrelia burgdorferi	Boerlin *et al.* (1992)
Candida albicans	Caugant and Sandven (1993)
Clostridium botulinum	Altwegg and Hatheway (1988)
Entamoeba histolytica	Nozaki *et al.* (1990)
Escherichia coli	Achtman *et al.* (1986) Mariani-Kurkdjian *et al.* (1993)
Fusobacterium spp.	Gharbia and Shah (1990)
Gluconobacter spp.	Yamada and Akita (1984)
Haemophilus influenzae	Musser *et al.* (1988)
Legionella pneumophila	Selander *et al.* (1985)
Leishmania spp.	Arana *et al.* (1990)
Listeria monocytogenes	Piffaretti *et al.* (1989)
Listeria spp.	Boerlin *et al.* (1991)
Mycobacterium avium	Yakrus *et al.* (1992)
Mycobacterium spp.	Wasem *et al.* (1991)
Mycoplasma spp.	O'Brien *et al.* (1981)
Neisseria gonorrhoeae	De La Feunte and Vazquez (1992)
Neisseria meningitidis	Caugant *et al.* (1986)
Plasmodium falciparum	Freese and Markus (1990)
Prevotella intermedia	Shah and Gharbia (1992)
Proteus spp.	Kappos *et al.* (1992)
Pseudocercosporella herpotrichoides	Priestly *et al.* (1992)
Pseudomonas syringae	Denny *et al.* (1988)
Rhynchosporium secalis	Newman (1985)
Salmonella typhi	Franco *et al.* (1992)
Streptococcus pneumoniae	Sibold *et al.* (1992)
Streptococcus spp.	Gilmour *et al.* (1987)
Treponema hyodysenteriae	Lymbery *et al.* (1990)
Trypanosoma brucei	Truc *et al.* (1991)
Trypanosoma cruzi	Breniere *et al.* (1991)
Vibrio cholerae	Salles and Momen (1991)
Xanthomonas maltophilia	Schable *et al.* (1991)
Yersinia pestis	Hudson *et al.* (1976)
Yersinia ruckeri	Schill *et al.* (1988)

PAGE can be performed in tube gels, more frequently a vertical slab-gel format is used. There are many slab electrophoresis systems available commercially so that both conventional, e.g. 20 × 20 cm, and mini-gels, e.g. 10 × 8 cm, can be prepared. The acrylamide solution is polymerized in a sealed mould comprising glass plates separated by spacers, whose thickness determines that of the gel (usually 0.75–2 mm). Use of the slab format permits the simultaneous analysis and side-by-side comparison of 20 or more samples on the same gel. PAGE can be performed using a buffer system which is uniform throughout the sample, gel and electrode chambers (a continuous buffer system) or, more commonly, using a multicomponent or discontinuous system which varies in pH and ion composition at different points. In the continuous system, the sample is loaded directly into the separating or resolving gel in which protein resolution occurs. In contrast, in the discontinuous system, samples are loaded into a low percentage (usually around 4%) acrylamide stacking gel cast on top of the resolving gel. The function of the stacking gel is essentially to concentrate components of the sample during the early stages of electrophoresis. This ensures that molecules enter the resolving gel in a sharp zone which enhances gel resolution.

PAGE can be performed either under non-denaturing conditions which maintain the native structure of protein molecules or, more commonly, under denaturing conditions where proteins are dissociated to their component polypeptides before electrophoresis. PAGE with non-denaturing conditions has not been used widely for typing or identification by conventional protein fingerprinting, as the resolution attained using this system may not always be sufficient for these purposes. However, this technique has been of value for typing of some microorganisms, and in particular some fungi. Recent applications include the differentiation of *Candida* spp. (Lehmann *et al.*, 1989a;b) and *Aspergillus* spp. (Matsuda *et al.*, 1992) by isoenzyme analysis on non-denaturing PAGE gels.

In contrast, protein profiles generated with denatured proteins have been used extensively for typing and identification, and by far the most widely used technique is sodium dodecylsulphate polyacrylamide gel electrophoresis (SDS-PAGE).

(c) SDS-PAGE

This method was described originally by Laemmli (1970). Minor modifications in the technique have since been made by other workers, but the basic principle of the technique remains the same.

To avoid effects of shape and native charge on electrophoretic mobility, and to increase gel resolution and reproducibility, proteins are denatured to linear polypeptides in the presence of an excess of the anionic detergent sodium dodecylsulphate (SDS) before electrophoresis. SDS binds to most polypeptides in a constant weight ratio that imparts a uniform negative charge density to the polypeptides and masks any intrinsic charge present. In gels of the correct concentration, polypeptides then migrate and separate strictly according to their size.

For many Gram-negative bacteria, and other microorganisms bounded by thin cell walls or lipid membranes, proteins can be easily solubilized and denatured in one step by resuspension and boiling for 5–10 min in SDS denaturation buffer (Table 5.1). For other organisms, e.g. staphylococci and fungi such as *C. albicans*, pretreatment of the microbial cells with wall-degrading enzymes, such as lysostaphin or xymolase, respectively, may be required to release proteins before denaturation in the presence of SDS and reducing agents. Soluble proteins released from microbes by physical methods are similarly denatured if analysis is to use SDS-PAGE.

To allow generation of reproducible profiles for comparative purposes, it is important that similar amounts of protein are loaded onto each lane of the gel. Typically, 25–100 μg of a complex protein mixture such as a whole-cell lysate would be loaded per lane, depending on the gel size and type of detection system to be used (see below). Protein concentrations of samples can be determined with a variety of commercially available protein assay kits.

Sample volumes are usually 10–75 μl and, where possible, identical volumes should be loaded on to each lane to ensure uniform lane width and band intensity. Overloading of lanes may distort the profile, while underloading may result in failure to detect minor components of the mixture. In addition to loading samples of the organisms of interest, molecular mass markers should also be electrophoresed on each gel. Single proteins, or mixtures of known molecular mass, are available commercially and their use not only allows the mass of unknown polypeptides present in samples to be determined, but also enables the reproducibility of electrophoresis runs to be compared.

The duration of electrophoresis varies with the equipment used, but run times of 3–4 h at 2–300 V constant voltage or 30 mA constant current are usual for a 20 × 20 cm gel format. A lower voltage (25–50 V) may be applied at the beginning of the run to allow polypeptides present in the sample to concentrate or stack in the stacking gel. With the minigel systems, run times of 45 min at 200 V constant voltage are often used.

Once electrophoresis is complete, polypeptides are fixed into the gel matrix and stained to allow visualisation of the polypeptide profile. Fixation and staining may be performed simultaneously, and the reagents used vary with different protocols. Polypeptides can be stained with general protein stains such as Coomassie Blue or PAGE Blue, or by a variety of silver-staining techniques (Oakley *et al.*, 1980; Sammons *et al.*, 1981) which offer greater sensitivity than general protein stains. Silver-staining techniques tend, however, to be more complex and, in some cases, prone to background problems. Some silver stains can also be applied to gels stained previously with general protein stains (e.g. Wray *et al.*, 1981). Polypeptide profiles generated using conventional and silver stains are not always identical, and may in fact be quite dissimilar as some polypeptides stain better with one stain type than with another. Once stained, gels can be stored in distilled water before analysis, though storage for extended periods of time may cause gels to become brittle and difficult to handle. To obtain a permanent record, gels should be photographed and, if required, can then be dried on to filter paper with a commercially-available gel dryer. Typical examples of SDS-PAGE analysis of whole cell and outer membrane protein profiles of *Escherichia coli* strains are shown in Fig. 5.2.

Where proteins have been radiolabelled before electrophoresis, gels are usually stained and dried as described above and then analysed by autoradiography. The profile of radiolabelled proteins is visualized following exposure to a suitable photograhic film (Hames, 1990).

Analysis of polypeptide profiles may be performed by direct visual comparison of adjacent sample lanes on the same gel. Major differences in profiles are detected readily by this method, particularly if the profile contains only a small number of major polypeptides, e.g. some bacterial outer-membrane preparations, and this approach may be sufficient for most typing purposes. The relative mobilities (rf) and molecular mass (kDa) of polypeptides of interest can be determined by construction of a standard curve with polypeptides of known mass electrophoresed on the same gel. Allocation of mass to individual proteins also facilitates inter-laboratory comparison of results. For some organisms, this approach has been extended by allocation of lettering or numbering schemes for particular groups of proteins. For example, a series of heat-modifiable outer-membrane proteins in meningococci and gonococci have been designated as Class 5 and PII proteins respectively (Tsai *et al.*, 1981; Stern and Meyer, 1987).

Where profiles are complex and more detailed comparisons are required, for example where overall similarities of profiles are needed for taxonomic studies, automated analysis is used (e.g. Hames, 1990; Wade *et al.*, 1990; Millership and Want, 1992). Stained gels can be scanned and images digitized using either a spectrophotometer with a suitable attachment or a dedicated instrument, such as a laser densitometer or image analysis system. Radiolabelled gels can also be scanned directly with a suitable detector (e.g. Tabaqchali *et al.*, 1987). These systems will generate a graphical representation of the scanned gel and some may also provide quantitative data. Analysis of the data generated can be performed by using a variety of mathematical models, and the usefulness of the data generated may be affected by the model chosen (Furst *et al.*, 1992).

Whole cell SDS-PAGE protein profiles have been used successfully in bacterial systematics as an alternative to conventional bioptyping methods for strain and species differentiation (Kersters and De Ley, 1975; Jackman, 1982, 1987). This technique has proven to be generally applicable to a wide range of bacteria, and overcomes the problem of having to use different sets of characteristics, e.g. biochemical tests, for diverse groups of organisms. Single concentration, rather than gradient gels, have given the best combination of protein resolution and reproducibility for this application. These studies have allowed a degree of taxonomic discrimination similar to that reported for DNA–DNA hybridization analysis (Jackman, 1987), and have generally used numerical taxonomic methods based on densitometer traces of gels for data generation (Feltham and Sneath, 1979; Jackman *et al.*, 1983).

A natural progression of this type of study is the use of protein profiles to create a data base against which fingerprints generated from unknown strains can be compared, forming the basis of an identification system (Kersters and De Ley, 1975). This approach has been proposed as a possible means of identifying bacteria in clinical laboratories

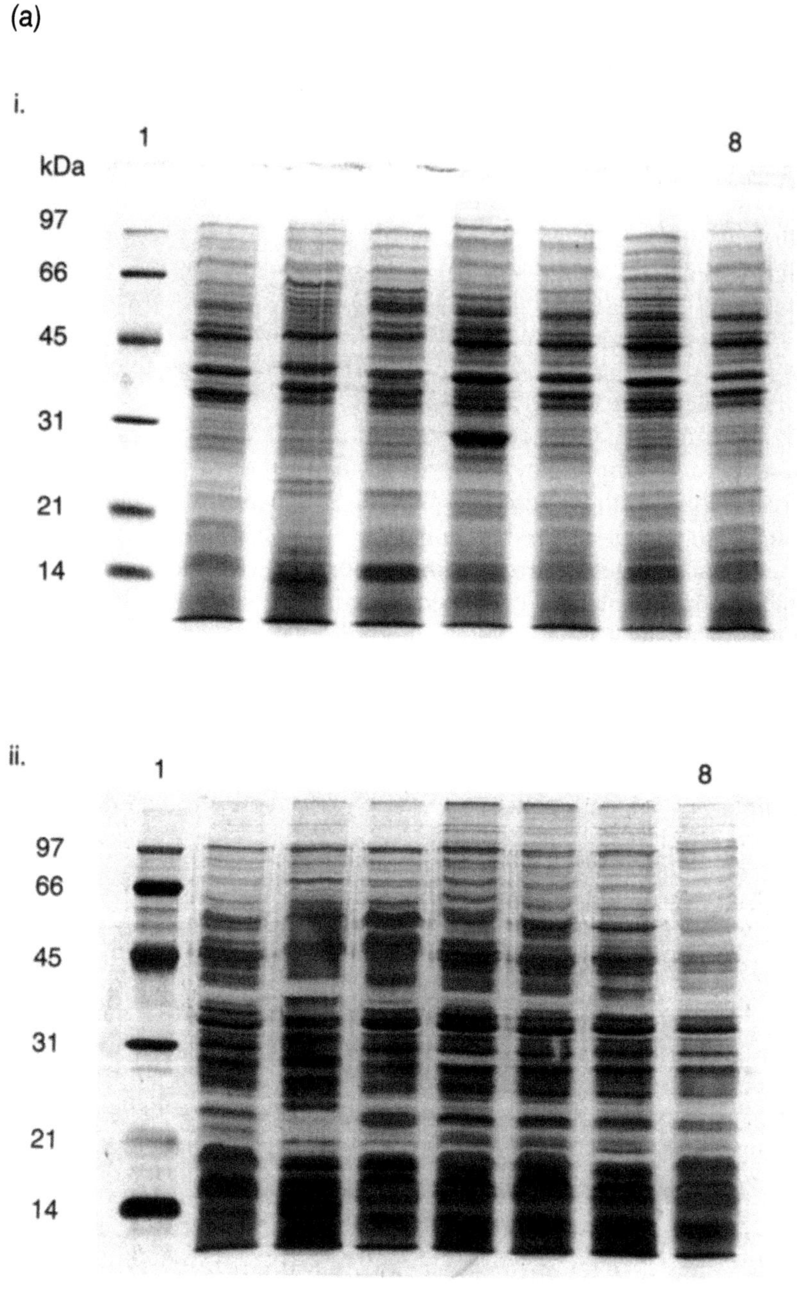

Figure 5.2 (a) Example of whole-cell protein profiles analysed by SDS-PAGE for seven *E. coli* strains. Molecular mass markers are shown in lane 1. (i) PAGE Blue 83-stained gel; (ii) the same gel restained with silver by the method of Wray *et al.* (1981).

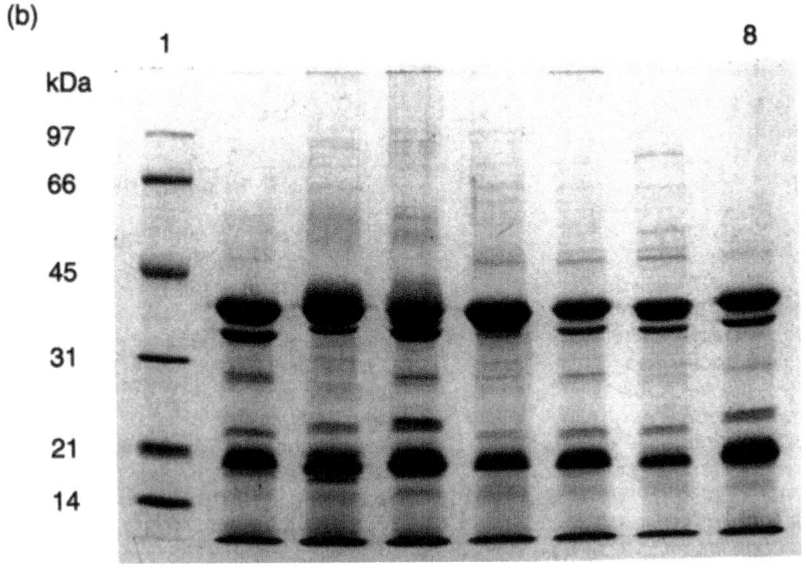

Figure 5.2 (b) Example of outer-membrane protein profiles of the *E. coli* strains shown in (a) (gel stained with PAGE Blue 83).

(Tabaqchali *et al.*, 1987). These workers used ^{35}S-labelled bacteria and an automatic scanning and analysis system to differentiate a wide range of Gram-positive and Gram-negative bacteria by means of whole-cell protein profiles generated by a technique called Radio-PAGE. Although this technique has been, and continues to be, used as a research tool for strain typing (Tabaqchali *et al.*, 1986; Stephenson & Tabaqchali, 1986; Clayton *et al.*, 1991), it has not yet found widespread application in routine diagnostic microbiology. Identification of bacteria to species level by conventional SDS-PAGE, e.g. *Mycobacterium* spp. (De Jong *et al.*, 1991), has also been reported.

In addition to its use in taxonomy and identification, SDS-PAGE has been used widely for typing and subtyping of microorganisms, particularly in epidemiological studies. Although the technique may, on its own, allow strain differentiation (e.g. Millership and Want, 1992; Walpita *et al.*, 1992), it has often been used in combination with other typing schemes such as serotyping, biotyping or MEE (e.g. Sibold *et al.*, 1992). Outer-membrane protein typing has proven particularly useful in epidemiological studies of meningitis caused by both *Neisseria meningitidis* (Achtman *et al.*, 1991) and *H. influenzae* (Barenkamp *et al.*, 1981). Typing schemes based on bacterial periplasmic proteins (Gargallo-Viola and Lopez, 1990) and penicillin binding proteins (Markiewicz and Tomasz, 1989; Sibold *et al.*, 1992) have also been reported. Other examples of the uses of SDS-PAGE for typing, taxonomic and epidemiological studies are listed in Table 5.4.

Table 5.4 Examples of the use of protein profiles produced by SDS-PAGE for typing and identification of microorganisms

Organism	Reference
Aeromonas spp.	Millership and Want (1989)
Bacteroides spp.	Poxton and Brown (1979)
	Dinsmoor *et al.* (1989)
Borrelia burgdorferi	Adam *et al.* (1991)
Campylobacter pyloridis	Megraud *et al.* (1985)
Campylobacter spp.	Dunn *et al.* (1987)
Chlamydia trachomatis	Batteiger *et al.* (1985)
Citrobacter diversus	Klein *et al.* (1989)
Coagulase-negative staphylococci	Stephenson and Tabaqchali (1986)
Corynebacterium group JK	Jackman and Pelczynska (1986)
Eubacterium spp.	Wade *et al.* (1990)
Haemophilus influenzae	Bruce and Jordens (1991)
Haemophilus influenzae type b	Barenkamp *et al.* (1981)
Lactococcus spp.	Elliott *et al.* (1991)
Morganella morgani	Senior and Voros (1990)
Mycobacterium tuberculosis	Millership and Want (1992)
Mycoplasma arthritidis	Washburn and Hirsch (1990)
Neisseria gonorrhoeae	Klempel and Clark (1988)
Neisseria meningitidis	Mocca and Frasch (1982)
Prevotella spp.	Yousefi-Mashouf *et al.* (1992)
Proteus spp.	Kappos *et al.* (1992)
Pseudomonas aeruginosa	Kelly *et al.* (1990)
Pseudomonas paucimobilis	Owen and Jackman (1982)
Rickettsiae	Beati *et al.* (1992)
Salmonella typhi	Franco *et al.* (1992)
Serratia marcescens	Gargollo-Viola and Lopez (1990)
Spiroplasmas	Mouches *et al.* (1979)
Staphylococcus aureus	Gaston *et al.* (1988)
Streptococcus pneumoniae	Sibold *et al.* (1992)
Ureoplasmas	Howard *et al.* (1981)

In some cases, SDS-PAGE profiles may not allow adequate discrimination to be achieved between strains. Profiles may be too complex, and significant diversity among strains of a given species may mean that this technique is unsuitable for identification purposes (e.g. Calhoon *et al.*, 1983). Alternatively, protein profiles of closely related strains within a serotype or subspecies may be identical, or show only minor differences which may not allow reproducible discrimination (e.g. Thornburg and Baseman, 1983). In some instances, even where protein differences that could allow strain differentiation exist, the technical limitations of SDS-PAGE may not allow sufficient resolution of proteins for typing purposes. In particular, proteins or polypeptides of identical or similar molecular mass may co-migrate during electrophoresis and produce a single band on the stained gel. The problem of co-migrating bands may be overcome if suitable antibodies are available (Chapter 6) to allow differentiation, but where these reagents are unavailable, the technique of two-dimensional SDS-PAGE (2D-PAGE) may be useful.

(c) Two-dimensional SDS polyacrylamide gel electrophoresis (2D-PAGE)

In contrast to one-dimensional SDS-PAGE, which separates polypeptides solely on the basis of molecular mass, 2D-PAGE is performed in two stages and uses both charge and mass to effect protein separation. Proteins are solubilized under conditions which preserve their native charge, and are first separated by isoelectric focusing. They are then subject to electrophoresis under denaturing conditions (SDS-PAGE) where resolution is dependent on molecular mass. Individual samples are run on each gel, and great care should therefore be taken to standardize gel preparation and electrophoresis conditions to enable comparisons to be made between gels. The original method of two-dimensional protein analysis was that described by O'Farrell (1975), and a more detailed description of the technique and its later modifications is provided by Rickwood *et al.* (1990).

Solubilization of proteins for isoelectric focusing should maintain their native charge, and usually involves the use of high concentrations of urea, non-ionic detergents and a reducing agent, in combination with a physical disruption method (Table 5.1). The urea used is ultra-pure and is usually deionized before use. Where these methods fail to produce adequate solubilization, SDS can be used, and high concentrations of urea and non-ionic detergent added subsequently to the sample. If possible, however, the use of SDS should be avoided as gel artefacts may result. In addition, to avoid modification of proteins during solubilization, extractions in the presence of high concentrations of urea should be performed at <60°C.

For 2D-PAGE, isoelectric focusing is performed routinely with low concentration (4%) polyacrylamide tube gels. The gel mix also contains urea, a suitable concentration of a non-ionic detergent to maintain protein solubility, and compounds called **ampholines**. Ampholine solutions contain a mixture of small, amphoteric compounds with closely spaced isoelectric points spanning a defined pH range. Ampholine solutions spanning several different pH ranges are available commercially. During electrophoresis, ampholines migrate in the electric field until they reach their pI, at which point they focus, thereby establishing a pH gradient along the tube gel. The tube gels are usually pre-electrophoresed to establish the pH gradient before loading the protein sample. Once loaded, proteins in the sample migrate in the electric field along the pH gradient until they reach their pI, at which point they focus. Again, care should be taken to load identical amounts of protein on to each tube gel to enable reproducible comparisons to be made. The amount of protein loaded will vary with sample type, apparatus used, and the detection method to be employed, but better resolution is generally obtained with small (c. 25 μl) sample volumes.

To increase reproducibility, electrophoresis conditions should be controlled carefully, for both the pre-electrophoresis and sample stages. Standardization of conditions, based on 'Volt-hours' (Volts $\times$ hours) is used frequently, and isoelectric focusing of the sample may take a few hours or overnight, depending on the conditions selected.

If it is necessary to know the pI of proteins of interest, marker proteins of known pI can be focused in duplicate tube gels. These can then be detected in the tube gels following staining, or after electrophoresis in the second dimension (see below). Alternatively, where large format tube gels are used, the gradient along the gel can be determined directly after focusing with a surface pH electrode. When the first dimension gels have been run, they may be applied directly to the second dimension gel or stored frozen for several weeks if necessary. The second dimension gels are conventional SDS-PAGE slab gels, which may be used with or without a stacking gel. To improve reproducibility of the second dimension gels, especially if gradient gels are used, it is preferable if several gels are poured at once from a single gel mix. Accessories for the simultaneous preparation of multiple slab gels are available commercially.

For the second dimension electrophoresis, the gel containing the focused proteins is extruded from the glass tube. Some protocols recommend equilibration of the tube gel with an SDS-containing buffer before application to the slab gel. If this step is omitted, the gel can be extruded directly on to the second dimension gel, which reduces the risk of damaging the tube gels during handling. Once positioned on the surface of the SDS-PAGE slab gel, the tube gel may be sealed in place by application of a molten agarose solution, but again this step is not essential. Electrophoresis in the second dimension is performed as described in section 5.2.2. (c), with molecular mass markers being run on one edge of the slab gel for reference purposes.

Once the second dimension electrophoresis is complete, gels are stained by an appropriate protocol. In contrast to one-dimensional SDS-PAGE, where polypeptides migrate as bands, polypeptides on 2D-PAGE gels are visualized as spots. Polypeptides of similar molecular mass, but different isoelectric points, are separated across the width of the gel. An example of a typical bacterial polypeptide profile generated

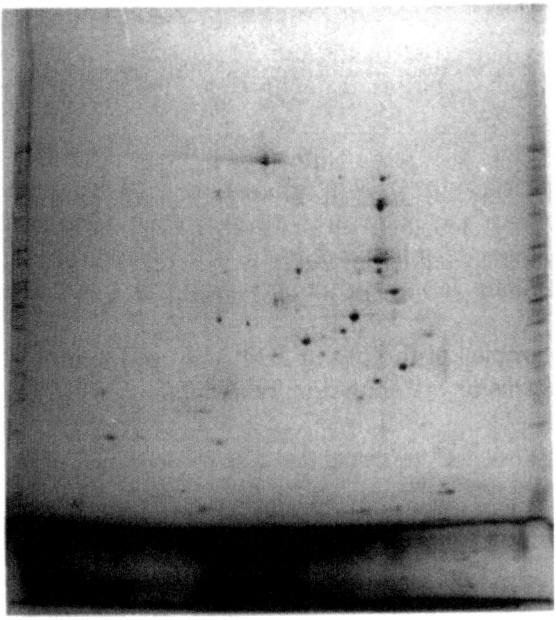

Figure 5.3 2D-PAGE profile generated with a whole-cell lysate. *Treponema pallidum* cells were solubilized as described in Table 5.1 for isoelectric focusing. Focused proteins were then separated in terms of mass by SDS-PAGE and stained with silver (Wray *et al.*, 1981).

by 2D-PAGE is shown in Fig. 5.3. Analysis of the polypeptide patterns produced is essentially as described for SDS-PAGE. Visual examination of the gels for the presence and relative positions of polypeptide spots allows direct comparisons to be made. Replicate gels are rarely identical because of variations in resolution and degree of stretching of the first dimension gels. However, patterns in replicates are usually sufficiently similar to reproducibly identify particular polypeptide spots. For detailed analysis of 2D-SDS profiles, computer-based analysis systems are available (Rickwood *et al.*, 1990; Cash, 1991).

Although the increased resolution offered by 2D-PAGE makes this technique an attractive one for typing and identification, the additional level of technical complexity, and limitations on the number of isolates that can be compared at one time, has prevented its widespread use for these purposes (Cash, 1991). Whole cell 2D-PAGE profiles have been used in bacterial taxonomy (Roberts *et al.*, 1980; Jackson *et al.*,

1985), while those based on outer-membrane proteins have been used for the detection of species-specific polypeptide differences between *Campylobacter* spp. (Dunn *et al.*, 1987). In some studies, use of 2D-PAGE has allowed detection of differences that were not readily apparent by conventional SDS-PAGE (Dunn *et al.*, 1987; Washburn and Hirsch, 1990).

2D-PAGE has also been applied to subtyping of viruses in order to identify markers of use in epidemiological studies (Cash, 1991). Extracts of radiolabelled, virus-infected cells separated by this technique have been used to differentiate variants within serotypes of poliovirus (Hamman *et al.*, 1978) and group B Coxsackie virus (Cash, 1989).

Other examples of the use of 2D-PAGE for identification and typing of microorganisms are shown in Table 5.5.

Table 5.5 Examples of the use of two-dimensional SDS-PAGE for strain differentiation of microorganisms

Organism	Reference
Aeromonas spp.	Cash (1991)
Aphthovirus	King *et al.* (1983)
Borrelia burgdorferi	Norris *et al.* (1992)
Campylobacter pylori	Dunn *et al.* (1989)
Campylobacter spp.	Dunn *et al.* (1987)
Chlamydia trachomatis	Batteiger *et al.* (1985)
Haemophilus influenzae	Cash (1991)
Mycoplasma arthritidis	Washburn and Hirsch (1990)
Mycoplasma mycoides	Rodwell and Rodwell (1978)
Neisseria gonorrhoeae	Klempel and Clark (1988)
Neisseria spp.	Jackson *et al.* (1985)
Plasmodium falciparum	Babiker *et al.* (1991)
Poliovirus	Hamman *et al.* (1978)
Propionibacterium spp.	Dekio *et al.* (1989)
Rhizobium spp.	Roberts *et al.* (1980)
Spiroplasmas	Mouches *et al.* (1979)

(e) Peptide mapping

Although peptide mapping has not been applied widely to typing of microorganisms, this technique has the potential to provide detailed information on the amino acid composition and relatedness of individual microbial proteins (Andrews, 1990). Selected proteins, common to all strains of interest, can be examined for structural variation to provide a peptide profile that allows their differentiation. Peptide mapping may be particularly useful where other separation methods, e.g.

SDS-PAGE, fail to discriminate between proteins, and where appropriate antisera are not available for immunological studies. This technique uses relatively small quantities of the protein of interest, and is performed using standard one or two-dimensional PAGE techniques.

The original peptide mapping method was described by Cleveland *et al.* (1977), and involves the treatment of proteins or polypeptides with proteases to produce partial or complete proteolytic digests. Chemical cleavage of proteins into fragments has also been used, but is generally more difficult to control than cleavage by means of proteases. The peptides produced can be separated and detected to produce a peptide map characteristic for that combination of protein, protease and incubation conditions. Different proteases cleave at different amino acid residues in a similar manner to that shown by restriction endonucleases in nucleic acid analysis. For example, *Staphylococcus aureus* V8 protease cleaves adjacent to glutamic acid or aspartic acid residues, whereas trypsin cleaves adjacent to arginine or lysine residues. The sensitivity of proteins to degradation is therefore dictated by their amino acid composition. Minor variations, e.g. single amino acid deletions or substitutions, in the composition of closely related proteins may result in the loss of a cleavage site and generation of a variant peptide map, allowing differentiation of the two molecules. Peptide mapping may also be used for the identification and isolation of suitable peptides for the generation of amino acid sequence data by automated sequencing techniques (e.g. O'Toole *et al.*, 1991).

Several methods of peptide mapping have been described (Andrews, 1990). Ideally, proteins to be mapped should be available in a pure form as described originally by Cleveland *et al.* (1977). This allows proteolysis of the pure protein to be performed in solution before peptide analysis. For many microbial proteins this may not, however, be feasible, and the initial stage of mapping usually involves separation of the components of a protein mixture, e.g. whole-cell lysate or membrane preparation, by PAGE, SDS-PAGE or isoelectric focusing. Following staining, the protein or polypeptide of interest is identified and excised from the gel with a scalpel and stored for later use. The amount of polypeptide or protein required for peptide mapping is dependent on the detection system to be used. If peptides are to be stained with Coomassie Blue, about 10 μg of protein is required, but this amount is reduced substantially if silver-staining or radiolabelling techniques are used.

In some protocols, the protein is eluted from the gel slice before treatment with protease, but recovery may be poor and, consequently, proteolysis is often performed directly on the gel slice in the sample well of the SDS-PAGE gel to be used for peptide analysis. The gel slice is pre-equilibrated in SDS-PAGE stacking gel buffer containing 0.1%

SDS before insertion into the sample well. An overlay of the same buffer containing 10% glycerol and an appropriate amount of protease is then applied. Electrophoresis is begun and results in the migration of the protease and protein present in the gel slice into the stacking gel. Proteolysis only occurs when the protease and protein come into contact, and a larger than usual stacking gel is therefore used to increase contact time. In some protocols, once the proteins have migrated towards the bottom of the stacking gel, the power supply is switched-off for about 30 min to facilitate proteolysis. Other workers use a constant, but low, voltage to ensure slow migration through the stacking gel. Once proteolysis has occurred, the power supply is switched-on and the voltage is increased to separate peptides in the resolving gel (usually 10–15% acrylamide). Better resolution may be obtained with gradient gels of between 10 and 20% acrylamide. Once electrophoresis is complete, peptides are detected by staining or autoradiography, as described in section 5.2.1. (c), and peptide patterns are compared for the different samples.

The use of peptide mapping allows the relatedness of proteins to be compared at the molecular level, and it is generally accepted that identical peptide profiles indicate that two proteins are identical (Andrews, 1990). Where peptide profiles show some common and some distinct features, a degree of relatedness is suggested. Occasionally, however, proteins with similar primary amino acid sequences may give different peptide profiles because of the relative accessibility of primary and secondary protease cleavage sites on the polypeptide. Accessibility is presumably influenced by secondary and tertiary protein structure. Discrepancies in the expected profiles can be confirmed or discounted by repeating the peptide mapping with other proteases or chemical cleavage methods, and comparing the profiles obtained.

Peptide mapping, in conjunction with SDS-PAGE, has been used to investigate the potential of variations in the molecular mass of protein A as a means of distinguishing between serotypes of *S. aureus* (Cheung *et al.*, 1987). Protein A variants that were not clearly distinguishable by SDS-PAGE could be differentiated by peptide mapping. Other workers have used peptide mapping to investigate the structural relationships of antigenic proteins or glycoproteins present on microbial cell surfaces. Barbour *et al.* (1983) used peptide mapping to investigate structural differences in conserved and serotype-specific membrane proteins of *Borrelia hermsii*. Similarly, Kutner *et al.* (1991) used peptide mapping to investigate the relationships between surface glycoproteins of *Leishmania* spp. in an attempt to identify antigens of diagnostic use. Peptide mapping has also been used to investigate structural changes in outer-membrane proteins of *Neisseria gonorrhoeae* in relation to changes in serum resistance (Judd *et al.*, 1987).

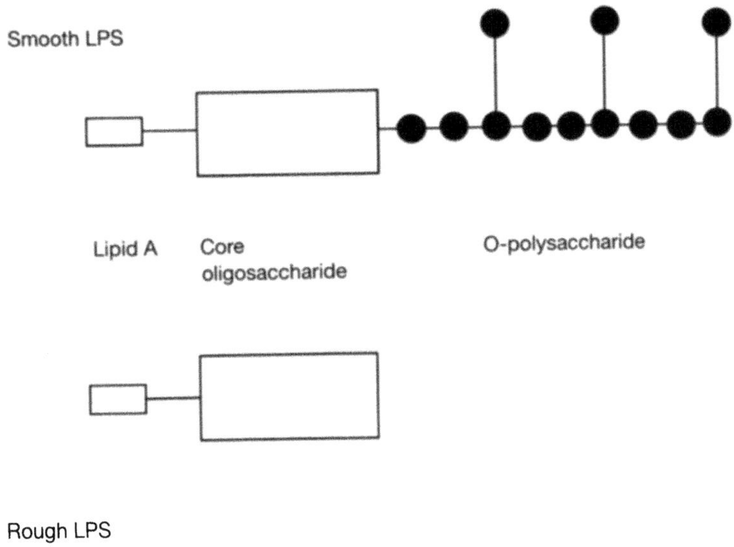

Figure 5.4 Schematic representation of the generalized structure of lipopolysaccharide.

5.3 LIPOPOLYSACCHARIDE TYPING SCHEMES

5.3.1 INTRODUCTION – STRUCTURE OF LIPOPOLYSACCHARIDES

Lipopolysaccharides (LPS) are components of the outer membrane of Gram-negative bacteria, and their application to typing and strain differentiation is, therefore, limited to this group of organisms. LPS is sometimes referred to as **endotoxin**, and is implicated as a virulence determinant in many pathogenic bacteria (Poxton and Arbuthnott, 1991).

LPS is a complex molecule consisting of two main components – a hydrophobic glycolipid called **lipid A**, and a hydrophilic polysaccharide component. Within this basic conserved structure, a significant level of structural and antigenic variability is apparent, permitting LPS analysis to be used as the basis of typing schemes.

The generalized structure of LPS molecules, based on current understanding, is shown in Figure 5.4. It should be noted, however, that although the structures of LPS isolated from different bacteria have been analysed in some depth (Brade *et al.*, 1988), it is generally accepted that, even within single molecules, micro-heterogeneity may exist, and more detailed analysis is still required to address this question.

Lipid A is responsible for the toxic activity of endotoxin, and its basic structure is more or less conserved among all LPS molecules which show toxic activity (LPS from some bacteria is weakly or non-toxic).

Minor variations occur in the length of the fatty acid residues linked to the disaccharide backbone of Lipid A, and in the substitutions which occur in the fatty acid and phosphoryl residues. Linked to Lipid A is a short core oligosaccharide, which comprises 11 or more monosaccharides and contains two unusual sugars, 3-deoxy-D-manno-2-octulosonic acid (KDO) and L-glycero or D-glycero-D-mannoheptose. This inner region of the core is broadly conserved over a wide range of bacterial species. The outer core – which contains more conventional sugars, including galactose, glucosamine and glucose – shows a limited degree of variability both between and within specieis. *Escherichia coli,* for example, exhibits several different LPS core types, whereas core structure is conserved among all *Salmonella* strains.

The O-polysaccharide components linked to the core constitute the O or somatic antigen of many Gram-negative bacteria and, in contrast to the core region, show an almost infinite degree of variation related to changes in the composition and arrangement of the sugar residues present. This variation is reflected in the large number of O serotypes of organisms such as *E.coli* and *Salmonella,* and forms the basis of a common and important method for differentiating and typing these and other Gram-negative bacteria (Chapter 6). O-polysaccharides exhibit marked variation in chain length and, within a single strain, chain length may vary considerably from nil to tens of repeated units. In addition, O-polysaccharide is not present in all types of LPS. LPS containing the O-polysaccharide is referred to as smooth. In contrast, some mutants produce LPS which lacks the O-polysaccharide, which is then referred to as rough LPS (Figure 5.4). In some organisms, e.g. *Neisseria* and *Haemophilus,* the LPS is naturally rough and is referred to as lipo-oligosaccharide or LOS.

The marked structural heterogeneity among LPS molecules of different Gram-negative bacteria also affects their physical properties, in particular molecular mass, allowing their differentiation by electrophoretic means. LPS and LOS can be extracted from bacteria and analysed by PAGE or SDS-PAGE to produce a fingerprint in a manner analagous to that used for proteins (Jann *et al.,* 1975).

5.3.2. METHODS USED IN LIPOPOLYSACCHARIDE PROFILE ANALYSIS

(a) Maintenance and growth of bacteria

For some bacteria that produce smooth LPS, repeated laboratory subculture may result in changes in the O-polysaccharide, and emergence of rough variants. Strains to be used for LPS typing should, therefore, be stored frozen and subcultured a minimal number of times before use. In addition, since LPS expression is a phenotypic character, conditions for the growth of bacteria should be standardized carefully. In particular, attention should be paid to standardization of the carbohydrate composition of the medium since marked variations may influence the sugar composition of the oligosaccharide core and O-polysaccharide side chains.

(b) Sample preparation

Traditionally, LPS has been extracted from bacteria before analysis. Smooth LPS can be extracted with hot aqueous phenol (Westphal and Jann, 1965), and rough LPS by treatment of bacteria with phenol, chloroform and petroleum spirit (Galanos *et al.*, 1969). These methods require large quantities of lyophilized bacteria and the use of hot phenol, which poses a potential safety risk. More recently, a rapid, non-solvent based method which uses very small quantities of bacteria, and which can be applied to both rough and smooth LPS and LOS, was described (Hitchcock and Brown, 1983). Bacterial lysates, prepared by boiling of bacteria in SDS-PAGE sample buffer, are treated with proteinase K to digest bacterial proteins, leaving the protease-resistant LPS in solution. For some organisms, clearer LPS profiles may be obtained with protease-treated outer-membrane preparations (Hancock and Poxton, 1988) which can be readily isolated by micro-methods such as that described by Carlone *et al.* (1986).

(c) Electrophoresis

LPS preparations are electrophoresed on 15% polyacrylamide gels as described in section 5.2.2. (b). Some workers omit SDS from the stacking and resolving gel mixes, but others report satisfactory resolution in the presence of SDS. Around 5 μg of LPS is applied per lane, following quantitation as described by Dubois *et al.* (1956). If LPS samples have been stored frozen, they should be heated before gel loading and electrophoresis to disperse any LPS aggregates that may have formed, since these may affect the profile obtained (Rimler, 1990).

Separated LPS molecules are detected by silver-staining (e.g. Tsai and Frasch, 1982). A method for detecting LPS by staining with ethidium bromide has been described by Kido *et al.* (1990).

An example of the type of profile generated for *E. coli* by this technique is shown in Figure 5.5.

(d) Analysis of LPS profiles

Lipopolysaccharide profiles of each strain are compared visually side-by-side on the same gel. The basic structure of the LPS molecules, i.e.

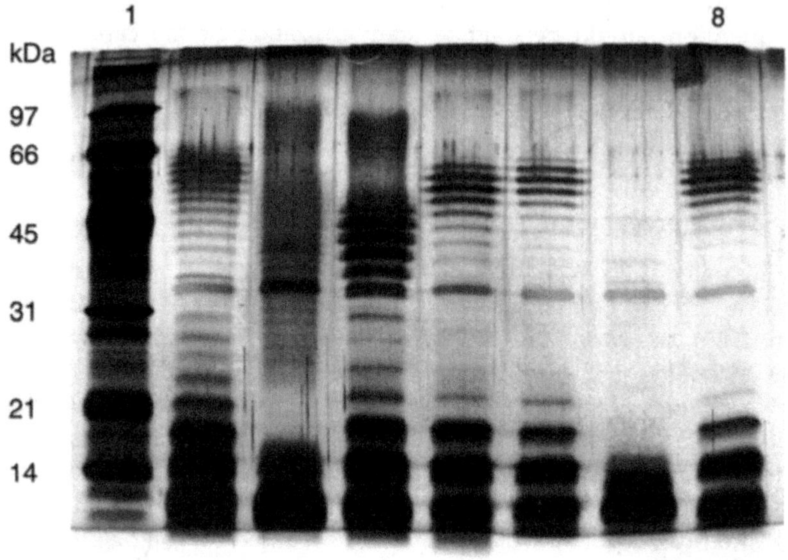

Figure 5.5 Lipopolysaccharide profiles generated for seven strains of *E.coli*. Outer-membrane preparations, prepared by the method of Carlone *et al.* (1986), were digested with proteinase K and electrophoresed on a 10% SDS-PAGE gel. SDS was omitted from the stacking and running gel buffers. The gel was stained by the method of Tsai and Frasch (1982). Molecular mass markers are shown in lane 1. An example of rough LPS is shown in lane 7.

rough or smooth, and the presence and molecular mass of stained bands relative to protein markers run on the same gel are recorded. In some cases, individual bands in some areas of the gel may not resolve, producing a smear rather than a discreet banding pattern. To assist with detailed interpretation of results, line drawings of the patterns may be made to improve clarity (e.g. Garin-Bastuji *et al.*, 1990).

(e) Applications of LPS profiles to typing

Although the major use of LPS for typing and identification has been in serotyping (Chapter 6), direct visualization of LPS banding patterns offers a rapid alternative method for differentiation of species and strains of a number of Gram-negative bacteria. For example, a variety of different serotypes of *Salmonella* showed distinct banding patterns on silver-stained gels (Luk and Lindberg, 1991). Similar observations were made with different serotypes of *Pasteurella multocida* (Rimler, 1990). LPS profiles have also been used to investigate inter-strain variation in both *Bordetella pertussis* (Peppler, 1984) and *H. influenzae* (Inzana, 1983).

Examples of other bacteria where comparative analysis of LPS profiles has been used for strain comparisons are shown in Table 5.6.

Table 5.6 Examples of organisms where LPS profiles have been investigated as a means of strain differentiation

Organism	Reference
Bacteroides fragilis	Weintraub *et al.* (1985)
Bordetella pertussis	Peppler (1984)
Brucella spp.	Garin-Bastuji *et al.* (1990)
Campylobacter pyloridis	Perez-Perez and Blaser (1987)
Campylobacter spp.	Perez-Perez and Blaser (1985)
Neisseria gonorrhoeae	Parr and Bryan (1984)
Neisseria meningitidis	Achtman *et al.* (1992)
Pasteurella haemolytica	Lacroix *et al.* (1993)
Pasteurella multocida	Rimler (1990)
Pseudomonas aeruginosa	Kelly *et al.* (1990)
Salmonella spp.	Hitchcock and Brown (1983)

5.4 SUMMARY

The main attractions of most of the techniques described in this chapter are their relative simplicity and general applicability to a broad range of microorganisms. Their application to any chosen organism requires only sufficient pure biomass for the analysis, and the development of a suitable solubilization protocol. The simple staining techniques used to generate fingerprints allow rapid direct comparisons to be made, and are not reliant on the availability of specific reagents such as antibodies or nucleic acid probes or primers. In most cases, large numbers of strains can be studied simultaneously and the data can be analysed by computer if necessary.

In many instances, these simple techniques provide sufficient discrimination between different isolates of closely related organisms to form the basis of typing and identification schemes. Where techniques such as SDS-PAGE fail to allow sufficient differentiation when used alone, they may be used in combination with other typing methods to improve discrimination. In particular, the use of specific antibodies for antigenic analysis offers an extra dimension to the use of both proteins and LPS for typing and identification. This topic will be covered in detail in the next chapter.

Although useful for strain typing, most of the techniques described give little or no information about the level of genotypic variation present in a group of closely related strains. The application of MEE over the past decade has, to some extent, filled this gap, and has provided an insight into the genetic diversity and clonal nature, particularly of bacterial populations. MEE also offers an alternative, highly discriminative method for basic typing applications, and has provided an extremely powerful tool for epidemiological studies of infectious diseases.

REFERENCES

Achtman, M., Heuzenroeder, M., Kusecek, B. *et al.* (1986) Clonal analysis of *Escherichia coli* 02:K1 isolated from diseased humans and animals. *Infection and Immunity*, **51**, 268–76.

Achtman, M., Wall., R.A., Bopp, M. *et al.* (1991) Variation in class 5 protein expression by serogroup A meningococci during a meningitis epidemic. *Journal of Infectious Diseases*, **164**, 375–82.

Achtman, M., Kusecek, B., Morelli, G. *et al.* (1992) A comparison of the variable antigens expressed by clone IV–1 and subgroup III of *Neisseria meningitidis* serogroup A. *Journal of Infectious Diseases*, **165**, 53–68.

Adam, T., Gassman, G.S., Rasiah, C. and Gobel, U.B. (1991) Phenotypic and genotypic analysis of *Borrelia burgdorferi* isolates from various sources. *Infection and Immunity*, **59**, 2579–85.

Altwegg, M. and Hatheway, C.L. (1988) Multilocus enzyme electrophoresis of *Clostridium argentinense* (*Clostridium botulinum* toxin type G) and phenotypically similar asaccharolytic *Clostridia*. *Journal of Clinical Microbiology*, **26**, 2447–9.

Andrews, A.T. (1990) Peptide mapping, in *Gel Electrophoresis of Proteins – A Practical Approach* (eds B.D. Hames and D. Rickwood), IRL Press, Oxford, pp. 301–20.

Arana, M., Evans, D.A., Zolessi, A. *et al.* (1990) Biochemical characterization of *Leishmania (Viannia) braziliensis* and *Leishmania (Viannia) peruviana* by isoenzyme electrophoresis. *Transactions of the Royal Society of Tropical Medicine and Hygiene*, **84**, 526–9.

Babiker, H.A., Creasey, A.M., Fenton, B. *et al.* (1991) Genetic diversity of *Plasmodium falciparum* in a village in eastern Sudan. 1. Diversity of enzymes, 2D-PAGE proteins and antigens. *Transactions of the Royal Society of Tropical Medicine and Hygiene*, **85**, 572–7.

Barbour, A.G., Barrera, O. and Judd, R.C. (1983) Structural analysis of the variable major proteins of *Borrelia hermsii*. *Journal of Experimental Medicine*, **158**, 2127–40.

Barenkamp, S.J., Munson, R.S. and Granoff, D.M. (1981) Subtyping isolates of *Haemophilus influenzae* type b by outer-membrane protein profiles. *Journal of Infectious Diseases*, **5**, 668–76.

Batteiger, B.E., Newhall, W.J. and Jones, R.B. (1985) Differences in outer membrane proteins of the lymphogranuloma venereum and trachoma biovars of *Chlamydia trachomatis*. *Infection and Immunity*, **50**, 488–94.

Beati, L., Finidori, J., Gilot, B. and Raoult, D. (1992) Comparison of serologic typing, sodium dodecyl sulfate polyacrylamide gel electrophoresis protein analysis, and genetic restriction fragment length polymorphisms analysis for identification of Rickettsiae: characterization of two new Rickettsial strains. *Journal of Clinical Microbiology*, **30**, 1922–30.

Boerlin, A.U., Rocourt, J. and Piffaretti, J.C. (1991) Taxonomy of the genus *Listeria* by using multilocus enzyme electrophoresis. *International Journal of Systematic Bacteriology*, **41**, 59–64.

Boerlin, P., Peter, O., Bretz, A.G. *et al.* (1992) Population genetic analysis of *Borrelia burgdorferi* isolates by multilocus enzyme electrophoresis. *Infection and Immunity,* **60**, 1677–83.

Brade, H., Brade, L. and Rietschel, E.T. (1988) Structure-activity relationships of bacterial lipopolysaccharides (endotoxins) – current and future aspects. *Zentralblatt für Bakteriologie und Hygiene,* **A268**, 151–79.

Breniere, S.F., Braquemond, P., Solari, J. *et al.* (1991) An isoenzyme study of naturally occurring clones of *Trypanosoma cruzi* isolated from both sides of the West Andes highland. *Transactions of the Royal Society of Tropical Medicine and Hygiene,* **85**, 64–6.

Bruce, K.D. and Jordens, J.Z. (1991) Characterization of noncapsulate *Haemophilus influenzae* by whole-cell polypeptide profiles, restriction endonuclease analysis, and rRNA gene restriction patterns. *Journal of Clinical Microbiology,* **29**, 291–6.

Calhoon, D.A., Myberry, W.R. and Slots, J. (1983) Cellular fatty acid and soluble protein profiles of oral fusobacteria. *Journal of Dental Research,* **62**, 1181–5.

Carlone, G.M., Thomas, M.L., Rumschlag, H.S. and Sottnek, F.O. (1986) Rapid microprocedure for isolating detergent-insoluble outer membrane proteins from *Haemophilus* species. *Journal of Clinical Microbiology,* **24**, 330–2.

Cash, P. (1988) Structural and antigenic variation of the structural protein VP3 in serotype–1 poliovirus isolated from vaccinees. *Canadian Journal of Microbiology,* **34**, 802–6.

Cash, P. (1989) Analysis of virus protein heterogeneity among Group-B coxsackie viruses using a mini two-dimensional gel electrophoresis system. *Electrophoresis,* **10**, 793–800.

Cash, P. (1991) The application of two-dimensional polyacrylamide gel electrophoresis to medical microbiology: molecular epidemiology of viruses and bacteria. *Electrophoresis,* **12**, 592–604.

Caugant, D.A. and Sandven, P. (1993) Epidemiological analysis of *Candida albicans* strains by multilocus enzyme electrophoresis. *Journal of Clinical Microbiology,* **31**, 215–20.

Caugant, D.A., Froholm, L.O., Bøvre, K. *et. al.* (1986) Intercontinental spread of a genetically distinctive complex of clones of *Neisseria meningitidis* causing epidemic disease. *Proceedings of the National Academy of Sciences of the USA,* **83**, 4927–31.

Cheung, A.L., Bayer, A.S., Peters, J. and Ward, J.L. (1987) Analysis by gel electrophoresis, western blot, and peptide mapping of protein A heterogeneity in *Staphylococcus aureus* strains. *Infection and Immunity,* **55**, 843–7.

Clayton, C.L., Kleanthous, H., Dent, J.C. *et al.* (1991) Evaluation of fingerprinting methods for identification of *Helicobacter pylori* strains. *European Journal of Clinical Microbiology and Infectious Diseases,* **10**, 1040–7.

Cleveland, D.W., Fischer, S.G., Kirschner, M.W. and Laemmli, U.K. (1977) Peptide mapping by limited proteolysis in sodium dodecyl sulfate and analysis by gel electrophoresis. *Journal of Biological Chemistry,* **252**, 1102–6.

Coakley, W.T., Bately, A.J. and Lloyd, D. (1977) Disruption of micro-organisms. *Advances in Microbial Physiology,* **16**, 279–341.

Davies, R.L., Porter, R., Coote, J.G. *et al.* (1992) Outer membrane protein and lipopolysaccharide variation in *Pasteurella haemolytica* serotype A1 under different growth conditions. *Journal of General Microbiology,* **138**, 909–22.

De Jong, A., Haentjen, A.H. and van der Zanden, A.G.M. (1991) A rapid method for identification of *Mycobacterium* species by polyacrylamide gel electrophoresis of soluble cell proteins. *Journal of Medical Microbiology*, **34**, 1–5.

Dekio, S., Hashimoto, K. and Makino, M. (1989) Two-dimensional gel electrophoresis of ribosomal proteins from *Propionibacterium acnes* and *granulosum*. *International Journal of Medical Microbiology*, **271**, 442–51.

De La Feunte, L. and Vazquez, J.A. (1992) Analysis of genetic variability of penicillinase non-producing *Neisseria gonorrhoeae* with different levels of resistance to penicillin. *Journal of Medical Microbiology*, **37**, 96–9.

Denny, T.P., Gilmour, M.N. and Selander, R.K. (1988) Genetic diversity and relationships of two pathovars of *Pseudomonas syringae*. *Journal of General Microbiology*, **134**, 1949–60.

Dinsmoor, M.J., Ebersole, J.E. and Gibbs, R.S. (1989) Protein banding patterns of the outer membrane-enriched fraction of *Bacteroides bivius*. *Journal of Clinical Microbiology*, **28**, 405–8.

Dubois, M., Gilles, K.A., Hamilton, J.K. and Rebers, P.A. (1956) Colorimetric method for the determination of sugars and related substances. *Analytical Biochemistry*, **28**, 350–6.

Dunn, B.E., Blaser, M.J. and Snyder, E.L. (1987) Two-dimensional gel electrophoresis and immunoblotting of *Campylobacter* outer membrane proteins. *Infection and Immunity*, **55**, 1564–72.

Dunn, B.E., Perez-Perez, G.I. and Blaser, M.J. (1989) Two-dimensional gel electrophoresis and immunoblotting of *Campylobacter pylori* proteins. *Infection and Immunity*, **57**, 1825–33.

Elliott, J.A., Collins, M.D., Pigott, N.E. and Facklam, R.R. (1991) Differentiation of *Lactococcus lactis* and *Lactococcus garvieae* from humans by comparison of whole-cell protein patterns. *Journal of Clinical Microbiology*, **29**, 2731–4.

Feltham, R.K.A. and Sneath, P.H.A. (1979) Quantitative comparison of electrophoretic traces of bacterial proteins. *Computers in Biomedical Research*, **12**, 247–63.

Franco, A., Gonzalez, C., Levine, O.S. *et al.* (1992) Further consideration of the clonal nature of *Salmonella typhi*: evaluation of molecular and clinical characteristics of strains from Indonesia and Peru. *Journal of Clinical Microbiology*, **30**, 2187–90.

Freese, J.A. and Markus, M.B. (1990) Enzyme typing of southern African isolates of *Plasmodium falciparum* by polyacrylamide gel electrophoresis. *Transactions of the Royal Society of Tropical Medicine and Hygiene*, **84**, 519–20.

Furst, V., Smith, I. and Holton, J. (1992) Computer analysis of ^{35}S labelled bacterial proteins. *Journal of Microbiological Methods*, **15**, 327–35.

Galanos, C., Luderitz, O. and Westphal, O. (1969) A new method for the extraction of R lipopolysaccharides. *European Journal of Biochemistry*, **9**, 245–9.

Gargollo-Viola, D. and Lopez, D. (1990) Numerical analysis of electrophoretic periplasmic protein patterns, a possible marker system for epidemiological studies. *Journal of Clinical Microbiology*, **28**, 136–9.

Garin-Bastuji, B., Bowden, R.A., Dubray, G. and Limet, J.N. (1990) Sodium dodecyl sulfate-polyacrylamide gel electrophoresis and immunoblotting analysis of smooth-lipopolysaccharide heterogeneity among *Brucella* biovars related to A and M specificities. *Journal of Clinical Microbiology*, **28**, 2169–74.

Gaston, M.A., Duff, P.S., Naidoo, J. *et al.* (1988) Evaluation of electrophoretic methods for typing methicillin-resistant *Staphylococcus aureus*. *Journal of Medical Microbiology*, **26**, 189–97.

Gharbia, S.E. and Shah, H.N. (1990) Identification of *Fusobacterium* species by the electrophoretic migration of glutamate dehydrogenase and 2-oxoglutarate reductase in relation to their DNA base composition and peptidoglycan dibasic amino acids. *Journal of Medical Microbiology*, **33**, 183–8.

Gilmour, M.N., Whittam, T.S., Kilian, M. and Selander, R.K. (1987) Genetic relationships among the oral streptococci. *Journal of Bacteriology*, **169**, 5247–57.

Hames, B.D. (1990) One-dimensional polyacrylamide gel electrophoresis, in *Gel Electrophoresis of Proteins – a Practical Approach* (eds B.D. Hames and D. Rickwood), IRL Press, Oxford, pp. 1–148.

Hamman, A., Reichel, C., Wiegers, K.J. and Drzeniek, R. (1978) Isoelectric points of polypeptides of standard poliovirus particles of different serological types and of empty capsids and dense particles of poliovirus type–1. *Journal of General Virology*, **38**, 567–70.

Hancock, I.C. and Poxton, I.R. (1988) Preparation of lipopolysaccharide and enterobacterial common antigen, in *Bacterial Cell Surface Techniques* (eds I.C. Hancock and I.R. Poxton), John Wiley, Chichester, pp. 90–6.

Hitchcock, P.J. and Brown, T.M. (1983) Morphological heterogeneity among *Salmonella* lipopolysaccharide chemotypes in silver-stained polyacrylamide gels. *Journal of Bacteriology*, **154**, 269–77.

Howard, C.J., Pocock, D.H. and Gourlay, R.N. (1981) Polyacrylamide gel electrophoretic comparison of the polypeptides from ureoplasmas isolated from cattle and humans. *International Journal of Systematic Bacteriology*, **32**, 125–31.

Hudson, B.W., Quan, T.J. and Bailey, R.E. (1976) Electrophoretic studies of the geographic distribution of *Yersinia pestis* protein variants. *International Journal of Systematic Bacteriology*, **26**, 1–16.

Inzana, T.J. (1983) Electrophoretic heterogeneity and interstrain variation of the lipopolysaccharide of *Haemophilus influenzae*. *Journal of Infectious Diseases*, **148**, 492–9.

Jackman, P.J.H. (1982) Classification of *Corynebacterium* species from axillary skin by numerical analysis of electrophoretic protein profiles. *Journal of Medical Microbiology*, **15**, 485–92.

Jackman, P.J.H. (1987) Microbial systematics based on electrophoretic whole-cell protein patterns, in *Methods in Microbiology*, vol. 19 (eds R. Colwell and R. Grigorova), Academic Press, London, pp. 209–25.

Jackman, P.J.H. and Pelczynska, S. (1986) Characterization of *Corynebacterium* Group JK by whole-cell protein patterns. *Journal of General Microbiology*, **132**, 1911–5.

Jackman, P.J.H., Feltham, R.K.A. and Sneath, P.H.A. (1983) A program in BASIC for numerical taxonomy of microorganisms based on electrophoretic protein patterns. *Microbios Letters*, **23**, 87–93.

Jackson, P., Thornley, M.J. and Thompson, R.J. (1985) A study by high resolution two-dimensional polyacrylamide gel electrophoresis of relationships between *Neisseria gonorrhoeae* and other bacteria. *Journal of General Microbiology*, **30**, 3189–201.

Jann, B., Reske, K. and Jann, K. (1975) Heterogeneity of lipopolysaccharide. Analysis of polysaccharide chain lengths by sodium dodecyl sulfate polyacrylamide gel electrophoresis. *European Journal of Biochemistry*, **60**, 239–46.

Judd, R.C., Tam, M. and Joiner, K. (1987) Characterization of protein I from serum-sensitive and serum-resistant transformants of *Neisseria gonorrhoeae. Infection and Immunity,* **55**, 273–6.

Kappos, T., John, M.A., Hussain, Z. and Valvano, M.A. (1992) Outer membrane protein profiles and multilocus enzyme electrophoresis analysis for differentiation of clinical isolates of *Proteus mirabilis* and *Proteus vulgaris. Journal of Clinical Microbiology,* **30**, 2632–7.

Kelly, N.M., MacDonald, M.H., Martin, N. *et al.* (1990) Comparison of the outer membrane protein and lipopolysaccharide profiles of mucoid and nonmucoid *Pseudomonas aeruginosa. Journal of Clinical Microbiology,* **28**, 2017–21.

Kersters, K. and De Ley, J. (1975) Identification and grouping of bacteria by numerical analysis of their electrophoretic protein patterns. *Journal of General Microbiology,* **87**, 333–42.

Kido, N., Ohta, M. and Kato, N. (1990) Detection of lipopolysaccharide by ethidium bromide staining after sodium dodecyl sulfate polyacrylamide gel electrophoresis. *Journal of Bacteriology,* **172**, 1145–7.

King, A.M.Q., McCahon, D., Nemman, J.W.I. *et al.* (1983) Electrofocusing structural and induced proteins of Apthovirus. *Current Topics in Microbiology and Immunology,* **104**, 219–33.

Klein, M.W., Mason, E.O. and Kaplan, S.L. (1989) Epidemiological marker system for *Citrobacter diversus* using outer membrane protein profiles. *Journal of Clinical Microbiology,* **27**, 1793–6.

Klempel, K.W. and Clark, V.L. (1988) Multiple protein differences exist between *Neisseria gonnorhoeae* type 1 and type 4. *Infection and Immunity,* **56**, 808–14.

Kutner, S., Pellerin, P., Breniere, S.F. *et al.* (1991) Antigen specificity of the 72-kilodalton major surface glycoprotein of *Leishmania braziliensis braziliensis. Journal of Clinical Microbiology,* **29**, 595–9.

Lacroix, R.P., Duncan, J.R., Jenkins, R.P. *et al.* (1993) Structural and serological specificities of *Pasteurella haemolytica* lipopolysaccharides. *Infection and Immunity,* **61**, 170–81.

Laemmli, U.K. (1970) Cleavage of structural proteins during the assembly of the head of bacteriophage T4. *Nature,* **227**, 680–5.

Lehman, P.F., Bin-Hsiao, C. and Salkin, I.F. (1989a) Protein and enzyme electrophoresis profiles of selected *Candida* species. *Journal of Clinical Microbiology,* **27**, 400–4.

Lehman, P.F., Kemker, B.J., Bin-Hsiao, C. and Dev, S. (1989b) Isoenzyme biotypes of *Candida* species. *Journal of Clinical Microbiology,* **27**, 2514–21.

Lehman, P.F., Chi-Wu, L. and Mackenzie, D.W.R. (1991) Isoenzyme changes in *Candida albicans* during domestication. *Journal of Clinical Microbiology,* **29**, 2623–5.

Luk, J.M.C. and Lindberg, A.A. (1991) Anti-*Salmonella* lipopolysaccharide monoclonal antibodies: characterization of *Salmonella* BO-, CO-, DO, and EO-specific clones and their diagnostic usefulness. *Journal of Clinical Microbiology,* **29**, 2424–33.

Lymbery, A.J., Hampson, D.J., Hopkins, R.M. *et al.* (1990) Multilocus enzyme electrophoresis for identification and typing of *Treponema hyodysenteriae* and related spirochetes. *Veterinary Microbiology,* **22**, 89–99.

Mariani-Kurkdjian, P., Denamur, E., Milon, A. *et al.* (1993) Identification of a clone of *Escherichia coli* 0103:H2 as a potential agent of hemolytic-uremic syndrome in France. *Journal of Clinical Microbiology*, **31**, 296–301.

Markiewicz, Z. and Tomasz, A. (1989) Variation in penicillin-binding protein patterns of penicillin-resistant clinical isolates of pneumococci. *Journal of Clinical Microbiology*, **27**, 405–10.

Matsuda, H., Kohno, S., Maesaki, S. *et al.* (1992) Application of ubiquinone systems and electrophoretic comparison of enzymes to identification of clinical isolates of *Aspergillus fumigatus* and several other species of *Aspergillus*. *Journal of Clinical Microbiology*, **30**, 1999–2005.

Megraud, F., Bonnet, F., Garnier, M. and Lamouliatte, H. (1985) Characterization of *Campylobacter pyloridis* by culture, enzymatic profile and protein content. *Journal of Clinical Microbiology*, **22**, 1007–10.

Millership, S.E. and Want, S.V. (1989) Typing of *Aeromonas* species by protein fingerprinting: Comparison of radiolabelling and silver staining for visualising proteins. *Journal of Medical Microbiology*, **29**, 29–32.

Millership, S.E. and Want, S.E. (1992) Whole-cell protein electrophoresis for typing *Mycobacterium tuberculosis*. *Journal of Clinical Microbiology*, **30**, 2784–7.

Mocca, L.F. and Frasch, C.E. (1982) Sodium dodecyl sulphate polyacrylamide gel typing system for characterization of *Neisseria meningitidis* isolates. *Journal of Clinical Microbiology*, **16**, 240–4.

Mouches, C., Vignault, J.G., Tully, T.G. *et al.* (1979) Characterisation of Spiroplasmas by one and two dimensional protein analysis on polyacrylamide gel slabs. *Current Microbiology*, **2**, 69–74.

Musser, J.M., Hewlett, E.L., Peppler, M.S. and Selander, R.K. (1986) Genetic diversity and relationships in populations of *Bordetella* spp. *Journal of Bacteriology*, **166**, 230–7.

Musser, J.M., Kroll, J.S., Moxon, E.R. and Selander, R.K. (1988) Clonal population structure of encapsulated *Haemophilus influenzae*. *Infection and Immunity*, **56**, 1837–45.

Newman, P.L. (1985) Variation amongst isozymes of *Rhynchosporium secalis*. *Plant Pathology*, **34**, 329–37.

Norris, S.J., Carter, C.J., Howell, J.K. and Barbour, A.G. (1992) Low-passage-associated proteins of *Borrelia burgdorferi* B31: characterization and molecular cloning of OspD, a surface-exposed, plasmid-encoded lipoprotein. *Infection and Immunity*, **60**, 4662–72.

Nozaki, I., da Silvaaca, E., Okuzawa, M. *et al.* (1990) Zymodemes of *Entamoeba histolytica* isolated in the Amazon and north-east Brazil. *Transactions of the Royal Society of Tropical Medicine and Hygiene*, **84**, 387–8.

Oakley, B.R., Kirsch, D.R. and Morris, N.R. (1980) A simplified ultrasensitive silver stain for detecting proteins in polyacrylamide gels. *Analytical Biochemistry*, **105**, 361–3.

O'Brien, S.J., Simonson, J.M., Grabowski, M.W. and Barile, M.F. (1981) Analysis of multiple isoenzyme expression among twenty-two species of *Mycoplasma* and *Acholeplasma*. *Journal of Bacteriology*, **146**, 222–32.

O'Farrell, P.H. (1975) High resolution of two-dimensional electrophoresis of proteins. *Journal of Biological Chemistry*, **250**, 4007–21.

O'Toole, P.W., Logan, S., Kostrzynska, M. *et al.* (1991) Isolation and biochemical and molecular characterization of a species-specific protein antigen from the gastric pathogen *Helicobacter pylori*. *Journal of Bacteriology*, **173**, 505–13.

Owen, R.J., & Jackman, P.J.H. (1982) The similarities between *Pseudomonas paucimobilis* and allied bacteria derived from analysis of deoxyribonucleic acids and electrophoretic protein patterns. *Journal of General Microbiology*, **128**, 2945–54.

Parr, T.R. and Bryan, L.E. (1984) Lipopolysaccharide banding patterns of *Neisseria meningitidis* and *Neisseria gonorrhoeae*. *Journal of Clinical Microbiology*, **19**, 558–60.

Peppler, M.S. (1984) Two physically and serologically distinct lipopolysaccharide profiles in strains of *Bordetella pertussis* and their phenotype variants. *Infection and Immunity*, **43**, 224–32.

Perez-Perez, G.I. and Blaser, M.J. (1985) Lipopolysaccharide characteristics of pathogenic *Campylobacters*. *Infection and Immunity*, **47**, 353–9.

Perez-Perez, G.L. and Blaser, M.J. (1987) Conservation and diversity of *Campylobacter pyloridis* major antigens. *Infection and Immunity*, **55**, 1256–63.

Pifferetti, J.C., Kressenbuch, H., Aeschbacher, M. *et al*. (1989) Genetic characterization of clones of the bacterium *Listeria monocytogenes* causing epidemic disease. *Proceedings of the National Academy of Sciences of the USA*, **86**, 3818–22.

Poxton, I.R. and Arbuthnott, J.P. (1991) Determinants of bacterial virulence, in *Topley and Wilson's Principles of Bacteriology Virology and Immunity*, 8th edn, vol. 1, (eds A.H. Linton and H.M. Dick), Edward Arnold, London, pp. 331–53.

Poxton I.R. and Brown, R. (1979) Sodium dodecyl sulfate polyacrylamide gel electrophoresis of cell-surface proteins as an aid to the identification of the *Bacteroides fragilis* group. *Journal of General Microbiology*, **112**, 211–7.

Priestly, R.A., Dewey, F.M., Nicholson, P. and Reyanoor, H.N. (1992) Comparison of isoenzyme and DNA markers for differentiating W-, R- and C-pathotypes of *Pseudocercosporella herpetotrichoides*. *Plant Pathology*, **41**, 591–9.

Rickwood, D., Chambers, A.A. and Spragg, S.P. (1990) Two-dimensional gel electrophoresis, in *Gel Electrophoresis of Proteins – a Practical Approach* (eds B.D. Hames and D. Rickwood), IRL Press, Oxford, pp. 217–72.

Rimler, R.B. (1990) Comparisons of *Pasteurella multocida* lipopolysaccharides by sodium dodecyl sulfate-polyacrylamide gel electrophoresis to determine relationship between Group B and E hemorrhagic septicaemia strains and serologically-related group A strains. *Journal of Clinical Microbiology*, **28**, 654–9.

Roberts, G.P., Leps, W.T., Silver, L.E. and Brill, W.J. (1980) Use of two-dimensional electrophoresis to identify and classify *Rhizobium* strains. *Applied and Environmental Microbiology*, **39**, 414–22.

Rodwell, A.W. and Rodwell, E.S. (1978) Relationships between strains of *Mycoplasma mycoides* sub spp. *mycoides* and *capri* studied by two-dimensional gel electrophoresis of cell proteins. *Journal of General Microbiology*, **109**, 259–63.

Salles, C.A. and Momen, H. (1991) Identification of *Vibrio cholerae* by enzyme electrophoresis. *Transactions of the Royal Society of Tropical Medicine and Hygiene*, **85**, 544–7.

Sammons, D.W., Adams, L.D. and Nishizawa, E.E. (1981) Ultrasensitive silver-based colour staining of polypeptides in polyacrylamide gels. *Electrophoresis*, **2**, 135–41.

Schable, B., Villarino, M.E., Favero, M.S. and Miller, J.M. (1991) Application of multilocus enzyme electrophoresis to epidemiologic investigations of *Xanthomonas maltophilia*. *Infection Control and Hospital Epidemiology*, **12**, 163–7.

Schill, W.B., Phelps, S.R. and Pyle, S.W. (1984) Multilocus electrophoretic assessment of the genetic structure and diversity of *Yersinia ruckeri*. *Applied and Environmental Microbiology*, **48**, 975–9.

Schwan T.G., Burgdorfer, W. and Garon, C.F. (1988) Changes in infectivity and plasmid profile of the Lyme disease spirochete, *Borrelia burgdorferi*, as a result of *in vitro* cultivation. *Infection and Immunity*, **56**, 1831–6.

Selander, R.K. and Musser, J.M. (1990) Population genetics of bacterial pathogenesis, in *The Bacteria, vol XI: Molecular Basis of Bacterial Pathogenesis*, (eds B.H. Inglewski and V.L. Clark), Academic Press, London, pp. 11–36.

Selander, R.K., McKinney, R.M., Whittam, T.S. *et al.* (1985) Genetic structure of populations of *Legionella pneumophila*. *Journal of Bacteriology*, **163**, 1021–37.

Selander, R.K., Caugant, D.A., Ochman, O. *et al.* (1986) Methods of multilocus enzyme electrophoresis for bacterial population genetics and systematics. *Applied and Environmental Microbiology*, **51**, 873–84.

Senior, B.W. and Voros, S. (1990) Protein profile typing – a new method of typing *Morganella morganii* strains. *Journal of Medical Microbiology*, **33**, 259–64.

Shah, H.N. and Gharbia, S.E. (1992) Multilocus enzyme electrophoretic analysis and DNA reassociation of strains designated *Prevotella intermedia*. *Letters in Applied Microbiology*, **15**, 8–11.

Sibold, C., Wang, J., Henrichsen, J. and Hakenbeck, R. (1992) Genetic relationships of penicillin-susceptible and -resistant *Streptococcus pneumoniae* strains isolated on different continents. *Infection and Immunity*, **60**, 4119–26.

Stephenson, J.R. and Tabaqchali, S. (1986) New methods for typing coagulase negative staphylococci. *Journal of Clinical Pathology*, **39**, 1271–5.

Stern, A. and Meyer, T.F. (1987) Common mechanism controlling phase and antigenic variation in pathogenic neisseriae. *Molecular Microbiology*, **1**, 5–12.

Tabaqchali, S., O'Farrell, S., Holland, D. and Silman, R. (1986) Methods for the typing of *Clostridium difficile* based on PAGE of ^{35}S-methionine-labelled proteins. *Journal of Clinical Microbiology*, **23**, 197–8.

Tabaqchali, S., Silman, R. and Holland, D. (1987) Automation in clinical microbiology: a new approach to identifying micro-organisms by automated pattern matching of proteins labelled with ^{35}S-methionine. *Journal of Clinical Pathology*, **40**, 1070–87.

Thornburg, R.W. and Baseman, J.B. (1983) Comparison of major protein antigens and protein profiles of *Treponema pallidum* and *Treponema pertenue*. *Infection and Immunity*, **42**, 623–7.

Tonolla, M., Demarta, A. and Peduzzi, R. (1991) Multilocus genetic-relationships between clinical and environmental *Aeromonas* strains. *FEMS Microbiology Letters*, **81**, 193–200.

Truc, P., Mathieudaude, F. and Tibayrenc, M. (1991) Multilocus isozyme identification of *Trypanosoma brucei* stocks isolated in central Africa – evidence for an animal reservoir of sleeping sickness in Congo. *Acta Tropica*, **49**, 127–35.

Tsai, C.M. and Frasch, C.E. (1982) A sensitive silver stain for detecting lipopolysaccharides in polyacrylamide gels. *Analytical Biochemistry*, **119**, 115–9.

Tsai, C.M., Frasch, C.E. and Mocca, L.F. (1981) Five structural classes of major outer membrane proteins in *Neisseria meningitidis*. *Journal of Bacteriology*, **146**, 69–78.

Wade, W.G., Slayne, M.A. and Aldred, M.J. (1990) Comparison of identification methods for oral asaccharolytic *Eubacterium* species. *Journal of Medical Microbiology*, **33**, 239–42.

Walpita, P., Mufson, M.A., Stanek, R.J. *et al.* (1992) Distinguishing between respiratory syncytial virus subgroups by protein profile analysis. *Journal of Clinical Microbiology*, **30**, 1030–2.

Wasem, C.F., McCarthy, C.M. and Murray, L.W. (1991) Multilocus enzyme electrophoresis analysis of the *Mycobacterium avium* complex and other Mycobacteria. *Journal of Clinical Microbiology*, **29**, 264–71.

Washburn, L.R. and Hirsch, S. (1990) Comparison of four *Mycoplasma arthritidis* strains by enzyme immunoassay, metabolism inhibition, one- and two-dimensional electrophoresis and immunoblotting. *Journal of Clinical Microbiology*, **28**, 1974–81.

Weintraub, A., Larsson, B.E. and Lindberg, A.A. (1985) Chemical and immunochemical analyses of *Bacteroides fragilis* lipopolysacchrides. *Infection and Immunity*, **49**, 197–201.

Westphal, O. and Jann, K. (1965) Bacterial lipopolysaccharide: extraction with phenol-water and further applications of the procedure. *Methods in Carbohydrate Chemistry*, **5**, 83–91.

Williams, R.A.D. and Shah, H.N. (1980) Enzyme patterns in bacterial classification and identification, in *Microbiological Classification and Identification* (eds M. Goodfellow and R.G. Board), Academic Press, London, pp. 299–318.

Wray, W., Bonlikas, T., Wray, V.P. and Hancock, R. (1981) Silver staining of proteins in polyacrylamide gels. *Analytical Biochemistry*, **118**, 197–203.

Yakrus, M.A., Reeves, M.W. and Hunter, S.B. (1992) Characterization of isolates of *Mycobacterium avium* serotypes 4 and 8 from patients with AIDS by multilocus enzyme electrophoresis. *Journal of Clinical Microbiology*, **30**, 1474–8.

Yamada, Y. and Akita, M. (1984) An electrophoretic comparison of enzymes in strains of *Gluconobacter* species. *Journal of General and Applied Microbiology*, **30**, 115–26.

Yousefi-Mashouf, R., Eley, A. and Duerden B.I. (1992) Characterization of non-pigmented species of the genus *Prevotella* by polyacrylamide gel electrophoresis. *Journal of Medical Microbiology*, **37**, 83–90.

Typing and identification of microorganisms with antibodies

6

6.1 INTRODUCTION

Although the generation of protein and LPS profiles allows adequate differentiation of organisms for some purposes, a much greater degree of discrimination may be obtained by using an additional property of these and other microbial molecules – their antigenicity. The development of serotyping schemes resulted from the early realization that specific antibodies produced by immunization of laboratory animals could be used to detect even subtle antigenic differences between microorganisms. Such differences were often not detectable by other methods, and the combination of sensitivity and specificity afforded by antibodies provided a major advance in typing methods. By using carefully characterized panels of antibodies, organisms sharing common major antigens are classified into serogroups, which are then further sub-divided into serotypes or serovars by using antibodies to antigens, or groups of antigens, whose production varies between different strains or isolates. The serotype finally allocated therefore usually reflects a combination of common and unique antigens of the organism, and can be used to differentiate and identify organisms at the genus and species level. Although all classes of microbial macromolecules are antigenic, antibodies to proteins, lipopolysaccharides and polysaccharides are used most widely. Different classes of molecule may be used to differentiate organisms at different levels. For example, *Neisseria meningitidis* strains are classified into serogroups by capsular polysaccharide typing, into serotypes on the basis of outer membrane protein profiles, and into sub-serotypes or immunotypes by LOS analysis (Gu *et al.*, 1992). By contrast, in *Chlamydia* spp., LPS constitutes a genus-specific antigen, and species are identified and sub-typed by outer-membrane protein analysis (Karimi *et al.*, 1989; Suchland and Stamm, 1991). The range of microorganisms for which serotyping schemes have been developed was shown earlier in Table 1.4.

For many of the earliest typing schemes, and for some still in use today, the detailed structure of many of the antigens recognized by antibodies used in serotyping remains poorly defined. Recent developments in the methods used to produce antibodies for typing purposes, combined with improved methods for antigen analysis and separation, have allowed the detailed structure of some important serotyping antigens to be examined. These studies should provide data to allow improvement and refinement of serotyping schemes, particularly for those problem organisms where these methods have proven unreliable or impracticable in the past.

This chapter will provide an overview of some of the traditional and newer methods that use antibodies for typing and identification of microorganisms. This is a constantly changing and expanding field, and more detailed accounts of the past and current methods applied to the serotyping of a range of medically-important microorganisms can be found in Parker and Collier (1990).

6.2 PRODUCTION OF ANTIBODIES FOR SEROTYPING AND IDENTIFICATION

All serotyping schemes rely on the availability of antibodies that will reproducibly detect antigenic differences between strains of the organisms of interest. The antigens used for strain differentiation should, therefore, be stably expressed and, if used for identification purposes, should be produced by all isolates of the species. Where such antibodies are used to identify specific organisms, for example in clinical microbiology laboratories, the specificity of the antibodies must also be tested to ensure that they do not cross-react with other organisms. Two broad classes of antibodies are used – polyclonal and monoclonal antibodies.

6.2.1 POLYCLONAL ANTIBODIES

Polyclonal antibodies are generated by immunization of laboratory animals, such as rabbits, guinea pigs, goats, sheep etc., with the antigen of interest. These were the only type of antibody available for early serotyping studies and are still used widely today. Since the protocols used for immunization may affect the antibody composition of the serum obtained, guidelines detailing standard immunization protocols have been established for many important pathogens (e.g. *Salmonella* spp.; Le Minor and Rhode, 1989).

By definition, polyclonal antisera contain a mixture of antibodies derived from many different B lymphocyte clones, each of which was stimulated by the immunizing antigen to produce a different antibody molecule. The antibodies produced may differ in class (e.g. IgM or IgG) and in isotype (e.g. IgG1, IgG2 etc.), and different antibodies may recognize alternative antigenic determinants or epitopes present at different sites on an individual antigen.

The antigen specificity of polyclonal sera for use in identification and typing can be achieved in two ways. Where an antigen of value in serotyping is available in purified form, it can be used directly for production of specific antibody. Immunization results in the generation of a variety of antibodies, all of which react with the immunizing antigen. Since these antibodies all react with a single antigen, they are referred to as monospecific. In most instances, however, antigens used for immunization are not available in purified form, and crude mixtures, e.g. whole microbial cells, are used for immunization. Sera produced by this method contain a variety of antibodies which recognize diverse epitopes on different microbial antigens, and are referred to as polyvalent sera. These sera may also contain antibodies to antigens common to a variety of strains or isolates, in addition to those of value for serological typing. For some applications, the presence of antibodies to common antigens poses no problem for strain differentiation. At other times, however, such antibodies are undesirable and have to be removed before the sera can be used for typing or identification purposes.

To remove antibodies of unwanted specificity, sera are absorbed repeatedly by incubation with antigen preparations that contain the common antigens, but lack the specific antigen of interest. Antibodies to the common antigens are removed by this process, and the resultant serum retains reactivity with those antigens of value for strain differentiation. The efficiency of the absorption process is checked subsequently by testing the specificity of the absorbed serum. This process of absorption and re-testing of serum specificity has to be repeated for each batch of serum used, and is consequently time-consuming and labour-intensive. Once sera capable of differentiating individual serogroups or serotypes are obtained, aliquots may be mixed to produce a single polyvalent serum that recognizes all known antigenic variants of that particular organism. These polyvalent sera are extremely useful for rapid presumptive identification of organisms before detailed serotyping is undertaken.

Although absorption to remove unwanted specificities works well with some organisms, particularly for many Gram-negative bacteria, success is reliant on selection of the appropriate antigens for absorption. Since, in some cases, the antigens involved in the generation of serotyping antisera are poorly defined, this approach may not always result in the expected antigen specificities being removed or retained. Even after absorption, cross-reactive antibodies which recognize other closely-related species may be retained. In addition, repeated absorption of sera may result in a reduction in the titre of specific antibody remaining, and this may limit the usefulness of sera in some typing methods. In addition, for some organisms, a proportion of isolates may be non-typable with the available antisera. Generation of additional antisera for newly identified serotypes involves repetition of animal immunization, absorption of sera, and re-testing to validate their usefulness. For species with large numbers of serotypes, or newly recognized pathogens, production of a comprehensive range of polyclonal antisera may take many years. However, polyclonal sera for serotyping some clinically important microorganisms are now readily available from commercial suppliers.

6.2.2. MONOCLONAL ANTIBODIES

The basic techniques for production of monoclonal antibodies were described by Kohler and Millstein (1975). This technology has found widespread application subsequently in all branches of the life sciences, and has proven particularly useful for identification and typing of many medically important microorganisms (Nowinski *et al.*, 1983; De Macario and Macario, 1983). Although the methods used in the production of monoclonals are, in some respects, more complex than those used for polyclonal antibodies, the use of monoclonals offers several advantages for many applications, including microbial typing and identification. A brief description of the technique will be given here; more detailed information on the principles and practice of monoclonal antibody production and use can be found in Goding (1983).

Monoclonal antibodies are produced by a two-stage process. Initially, mice or rats are immunized with the antigen of interest. This may be a purified molecule, but this is not essential and usually a crude antigen mixture is used. As with the generation of polyclonal sera, immunization leads to the formation of a diverse range of antibodies, each secreted by an individual B lymphocyte clone. Once immunization is complete, animals are sacrificed and the antibody-secreting B lymphocytes are extracted from the spleen or lymph nodes. These are then mixed with myeloma cells (immortal cells which will grow indefinitely in culture) under conditions that cause fusion of the cell membranes and production of a stable hybrid cell or **hybridoma**. These cells exhibit the combined phenotypes of both parent cells, and secrete antibody of the epitope specificity, class and isotype determined by the B lymphocyte, but which, unlike B lymphocytes, will grow indefinitely in culture.

The antibody produced is secreted into the medium used to grow the cells, and hybridomas secreting antibody to individual antigens present in the original immunizing antigen are selected by an appropriate screening technique. For individual antigens, hybridomas producing antibodies specific for different epitopes present on the antigen can be selected. Once hybridomas are selected, they are cloned by limiting dilution to ensure that each cell line is indeed derived from a single cell, i.e. is monoclonal. Stock cultures can then be stored frozen in liquid nitrogen until required. The process of immunization, hybridoma generation, and selection and cloning of cell lines to produce appropriate antibodies may take several months and requires expertise in cell culture techniques.

To produce a stock of antibody for use in typing or identification, hybridomas are grown in large quantities and the supernatant containing the antibody is harvested. If necessary, the antibody may be purified and concentrated before use. Alternatively, high-titre monoclonal antibody may be produced by growth of hybridoma cells in the peritoneal cavities of mice or rats. Growth of the cells results in the accumulation of fluid – an **ascites** – in the peritoneum. The ascitic fluid is harvested and may either be used without further treatment or fractionated to recover the antibodies.

As a consequence of the methods used to generate and select monoclonal antibodies, many of the potential problems associated with the use of polyclonal sera are avoided. Specific antibodies can be generated without the requirement for purified antigen or absorption of sera. However, cross-reactive monoclonals, which recognize structurally-related epitopes on otherwise unrelated antigens, may also be produced. Monoclonals should, therefore, also be screened for cross-reactivity with a range of organisms before being considered for use in the identification of microorganisms. These techniques also allow selection of antibodies reactive with individual epitopes on antigens, potentially increasing the level of discrimination that can be achieved. Where monoclonals are to be used for identification, it is therefore important that the epitope selected is conserved throughout isolates of the organism of interest. Finally, once generated, hybridomas offer a potentially limitless and uniform supply of a specific antibody without the requirement for repeated immunizations, thereby significantly reducing animal usage for antibody generation. Examples of organisms where monoclonal antibodies have been used in typing or identification are listed in Table 6.1. Some of these antibodies are available commercially, whereas the use of others is restricted to research laboratories.

Table 6.1 Examples of organisms for which monoclonal antibodies have been used for identification or strain differentiation

Organism	Reference
Bordetella pertussis	Gustafsson *et al.* (1988)
Borrelia burgdorferi	Barbour *et al.* (1985)
Borrelia hermsii	Schwan *et al.* (1992)
Bovine RSV	Baker *et al.* (1992)
Brucella spp.	Garin-Bastuji *et al.* (1990)
Chlamydia trachomatis	Barnes *et al.* (1985)
Cryptococcus neoformans	Hamilton *et al.* (1991a)
	Dromer *et al.* (1993)
Coxsackie B3 virus	Yagi *et al.* (1992)
Coxsackie B4 virus	Prabhakar *et al.* (1982)
Cryptosporidium spp.	Garcia *et al.* (1992)
Echovirus type 25	Peigue-Lafeuille *et al.* (1990)
Entamoeba histolytica	Gonzalez-Ruiz *et al.* (1992)
Equine rotavirus	Browning *et al.* (1992)
Francisella tularensis	Fulop *et al.* (1991)
Giardia spp.	Garcia *et al.* (1992)
Haemophilus ducreyi	Karim *et al.* (1989)
Haemophilus influenzae	Moller *et al.* (1992)
Helicobacter pylori	Drouet *et al.* (1991)
Herpes simplex virus	Lipson *et al.* (1991)
Human CMV	Boppana *et al.* (1992)
Human rotavirus	Cukor *et al.* (1984)
Human rotavirus VP4	Coulson (1993)
Human RSV	Anderson *et al.* (1985)
Histoplasma capsulatum	Hamilton *et al.* (1991b)
Legionella pneumophila	Joly and Winn (1984)
Legionella spp.	Steinmetz *et al.* (1991)
Leishmania spp.	Hanham *et al.* (1991)
Leptospira interrogans	Terpstra *et al.* (1985)
Neisseria gonorrhoeae	Tam *et al.* (1982)
	Knapp *et al.* (1984)
Neisseria meningitidis	Achtman *et al.* (1992)
	Gu *et al.* (1992)
Mycobacterium avium	Abe *et al.* (1989)
Nocardia spp.	Boiron *et al.* (1992)
Ovine lentivirus	Marcom *et al.* (1991)
Pasteurella haemolytica	Lacroix *et al.* (1993)
Pneumocystis carinii	Lundgren *et al.* (1992)
Porcine group C rotavirus	Ojeh *et al.* (1991)
Salmonella spp.	Luk and Lindberg (1991)
Shigella flexneri	Carlin *et al.* (1989)
Sporothrix schenckii	Hamilton *et al.* (1991c)
Staphylococcus aureus	Nelles *et al.* (1985)
Streptococcus pneumoniae	Hakenbeck *et al.* (1991)
Treponema pallidum	Hook *et al.* (1985)
Treponema pertenue	Noordhoek *et al.* (1990)
Vibrio cholerae	Gustafsson *et al.* (1982)

Despite these obvious advantages, there are also some potential problems with the use of monoclonals for typing and identification purposes. These include their variable and sometimes weak affinity, and occasional problems with the isolation of antibodies reactive with epitopes of value for strain differentiation. In addition, monoclonals that recognize conformational epitopes on intact antigen molecules may not react once the antigen has been denatured. The antibody class and isotype of a monoclonal may also limit its direct use in some techniques which involve agglutination or precipitation reactions. It is important, therefore, that the method used initially to select monoclonals is similar to that to be used for the final typing or identification technique.

6.3 TECHNIQUES USED IN SEROTYPING

6.3.1. TRADITIONAL METHODS

A variety of techniques such as slide and tube agglutination, co-agglutination, immunofluorescence and gel precipitation have been used for serotyping of microorganisms (Ørskov and Ørskov, 1984). Some of these techniques are still used widely in clinical microbiology laboratories, reflecting their continuing value for routine diagnostic applications.

(a) Agglutination

Historically, polyclonal sera were the only reagents available and were used in slide or tube agglutination tests with whole viable or killed bacteria as antigen. Specific antibodies in the sera bound to antigens on the bacterial surface to produce visible agglutination of the bacterial suspension. A modification of this technique, called **co-agglutination**, has also been used extensively. In this method, the typing antibody is first incubated with *Staphylococcus aureus* cells, where it binds to protein A molecules on the bacterial surface. Mixing of the antibody-coated *S. aureus* cells with the test organism then results in agglutination if the antibody and specific antigen combine.

Agglutination tests form the basis of current serotyping and identification schemes for many Gram-negative bacteria, including *Salmonella* spp. (Kaufmann, 1972) and *Escherichia coli* (Ørskov and Ørskov, 1984), and use antibodies specific for LPS (O), capsular (K) and flagellar (H) antigens. Similar methods have been applied to the typing of Gram-positive bacteria, such as *Streptococcus pneumoniae* (Lund and Henrichsen, 1978), where strain differentiation is based on surface proteins, and to the fungus *Candida albicans* (Hasenclever, 1965). More recently, serotyping assays relying on the same basic principle, but using antibodies coated on to latex beads to improve the ease of detecting agglutination reactions, have been developed for some bacteria (e.g. *Salmonella* spp.; Bouvet and JeanJean, 1992). Co-agglutination techniques are still used widely. For example, Danielsson *et al.* (1988) used co-agglutination with polyclonal sera for strain differentiation of *Helicobacter* (formerly *Campylobacter) pylori*.

Although agglutination tests have proven useful for typing many microorganisms, the level of discrimination that can be achieved with this technique varies markedly for different organisms. Although around 2000 different *Salmonella* serotypes can be defined by slide and tube agglutination (Kaufmann, 1972), similar methods can only identify two major antigenic groups (A and B) in *C. albicans* (Hasenclever, 1965). This naturally limits the usefulness of this technique for detailed strain differentiation of *C. albicans*, and other typing methods, such as immunoblotting (next section), may be of more value with this and some other organisms.

Slide agglutination tests have remained popular in routine diagnostic laboratories because they are simple, use small quantities of reagents, require no specialized equipment and, in conjunction with a limited number of additional phenotypic and biochemical tests, will allow presumptive identification of an organism to the genus or species level in a few minutes. The usefulness of these techniques is limited only by the availability and specificity of appropriate typing sera. As outlined above, the necessity to absorb sera for typing purposes may produce sera of low titre and variable specificity, particularly when complex antigens are involved. The use of monoclonal antibodies for typing and identification offers the potential to overcome these problems and produce truly specific reagents for these applications. This approach has recently been reported for *Shigella* spp. (Carlin *et al.*, 1989) and *Salmonella* spp. (Luk and Lindberg, 1991). The latter workers used monoclonals specific for natural or synthetic oligosaccharides present on the LPS from salmonellae of serogroups A–E to discriminate between a range of species. Monoclonals performed as well as conventional polyclonal sera, and eliminated cross-reactions with other closely related Enterobacteriaceae.

A problem with some monoclonals, particularly those of IgG isotypes, is that they fail to agglutinate particular antigens such as bacteria. The value of these monoclonals for routine typing or identification purposes can be enhanced by the co-agglutination technique. Coating of the monoclonal on to the surface of *S. aureus* increases its ability to cross-link antigens and cause agglutination (e.g. Carlin *et al.*, 1989).

(b) Immunofluorescence

Immunofluorescence has also been used widely for the typing and identification of a range of microorganisms. Again, this technique generally (though not exclusively) detects surface-associated antigens. Slide formats, with organisms fixed directly to the slides, are used for bacteria. For some viruses, and *Chlamydia*, infected tissue culture cells are used as the antigen. Recently, an immunofluorescence method for typing *Chlamydia trachomatis* directly in cell culture monolayers has been described (Suchland and Stamm, 1991). After fixation, antigen is incubated with specific antibody. This may be labelled directly with fluorescein to allow visualization of antibody (a direct immunofluorescence assay) or, alternatively, a fluorescein-labelled antibody conjugate is added in a second stage to detect bound antibody (an indirect immunofluorescence assay). Slides are then examined with an ultraviolet microscope, and the intensity of fluorescence, relative to appropriate controls, is scored on an arbitrary scale from − to +++++. Typical assays can be performed in 2–3 h. A more detailed description of immunofluorescence techniques is given by Edwards, Taylor and Tomlinson (1982).

Immunofluorescence with polyclonal sera has been used for serotyping a number of different microorganisms, but the technique does have some problems. These include non-specific reactions and high background levels. In addition, reading of results is subjective and, when large numbers of strains are involved, may be very time-consuming. Monoclonal antibodies have been used with immunofluorescence for typing some organisms, e.g. *Legionella pneumophila* (Joly and Winn, 1984), resulting in increased specificity of the reactions, and monoclonal antibodies are now used routinely for identification of legionellae in clinical specimens. This approach has been used for other organisms and, most recently, a typing system for the fungus *Cryptococcus neoformans*, based on an immunofluorescence assay using a monoclonal antibody, has been described (Dromer *et al.*, 1993). The problem of quantitating fluorescence to remove subjective evaluation of test results has been addressed, but although automated experimental systems for quantitation have been developed (Kaplan and Picciolo, 1989), these have not been applied widely to date.

6.3.2 IMMUNOBLOTTING

The traditional methods, described above, generally use a limited number of surface-associated microbial antigens to achieve strain differentiation. More detailed information of value for serotyping can be obtained by looking at a broader range of microbial antigens, including surface-associated antigens and those located inside the organism. In addition, the antigens used for traditional serotyping are generally analysed in their native form, limiting the analysis to epitopes present on the surface of the antigens. Denaturation of antigens may expose additional epitopes of value in serotyping. One of the techniques used most frequently for analysis of denatured antigens, and which has found widespread use in serotyping of microorganisms, is western blotting or immunoblotting.

The basic technique for immunoblotting was described by Towbin *et al.* (1979), and involves the electrophoretic transfer of proteins or LPS molecules – separated by SDS-PAGE – from polyacrylamide gels to a nitrocellulose or nylon membrane. Membranes are then incubated with a polyclonal serum or monoclonal antibody, after which antibody binding to antigens is detected by incubation with a tagged antibody conjugate and an appropriate detection system. Transfer of the antigens to the surface of the membrane increases their availability for antibody binding and significantly speeds-up the immunodetection process. A brief description of the technique will be given here; the interested reader is referred to Towbin and Gordon (1984) and Hames (1990) for detailed methodologies.

(a) Electroblotting of polypeptides and LPS

Samples are prepared and electrophoresed on SDS-PAGE gels as described in Chapter 5. Care should again be taken to ensure standardization of growth conditions etc. for the test strains to avoid any phenotypic variation which could lead to changes in antigen profiles. Since immunoblotting is a very sensitive technique, smaller quantities of protein are generally electrophoresed than are used for SDS-PAGE fingerprinting.

After electrophoresis, gels are usually pre-equilibrated in transfer buffer for about 30 min. A typical transfer buffer consists of 25 mM Tris, 192 mM glycine, 20% v/v methanol, pH 8.3. The methanol in the buffer prevents the gels swelling during electro-transfer and improves blot resolution. In addition, methanol promotes binding of proteins to some membranes. SDS (0.1%) may be added to the buffer to enhance transfer of some large polypeptides, but this may also reduce binding efficiency for some membrane types. Several different membranes are available commercially. Nitrocellulose membranes of 0.45 or 0.2 μm pore size were used originally, but have the disadvantage that they are flimsy and brittle when dry. Nylon membranes were introduced later and offered higher protein-binding capacity, but were sometimes prone to high background problems. Supported nitrocellulose membranes, which provide a compromise between pure nitrocellulose and nylon, are now available and these perform well in the authors' experience.

For the electro-transfer, the equilibrated gel is overlaid with a piece of the blotting membrane and sandwiched between pieces of filter paper soaked in the transfer buffer. Trapping of air bubbles between the gel and membrane or filter paper will prevent transfer and must be avoided. The transfer is performed using one of several commercially available transfer systems. These can be divided into two basic types – those where the gel and membrane are completely submerged in buffer during transfer (wet systems), and those where a minimal volume of buffer is used (semi-dry systems). Electrophoresis times and conditions for different systems vary, but electrophoresis at 300 mA constant current for 3 h, or 100 mA overnight, are common. Cooling may be required. Shorter electrophoresis times are used with semi-dry systems, and runs of 15–30 min at 10–15 V constant voltage, or 1 h at 15–25 V, have been used for mini- and standard-sized gels, respectively. With both types of system, several gels may be transferred at once, thereby reducing the time required to analyse large numbers of strains. Generally, transfer is efficient, but occasionally some large polypeptides (>100 kDa) may not transfer efficiently in the presence of methanol, and modification of the buffer formulation may be required. Combinations of high current or voltage and extended transfer times may result in small polypeptides and some LPS molecules passing straight through the membrane. Consequently, optimization of transfer conditions for individual organisms may require some preliminary experimentation. The efficiency of transfer of polypeptides can be checked by staining membranes with a general protein stain such as Naphthalene Blue 12B or Ponceau S. This also provides a complete profile of the polypeptides transferred, thereby allowing antigens of value in serotyping to be identified by comparison with the original gel. Transfer and staining of molecular mass markers also aids in identifying and characterizing useful antigens. Pre-stained markers are available commercially for this purpose.

For some applications, antigen preparations can be spotted directly onto transfer membranes. This technique, called **dot blotting**, removes the need for SDS-PAGE and electro-transfer, and therefore significantly speeds-up the whole process. The antigen is usually in a native form, and may be whole microbial cells, cell sonicate or purified antigen.

(b) Immunodetection

Before incubation with antibodies, membranes are first incubated for about 1 h in a blocking solution to saturate the remaining free protein-binding sites on the membrane. The blocking agents used are immunologically-inert proteins such as bovine serum albumin (BSA), gelatin or dried milk powder, though some detergents such as Tween-80 have also been used. This blocking step prevents the non-specific binding of antibody to the membrane and reduces background staining.

Membranes are then incubated with antibody diluted in a physiological buffer such as phosphate-buffered saline (PBS), which may be supplemented with BSA and detergent to further reduce non-specific antibody binding. A typical antibody dilution buffer is PBS containing BSA 0.1% w/v and Tween-20 0.1% v/v, pH 7.4. Tween-20 may cause leaching of some molecules from the membrane, and may be omitted from some applications. Antibody dilutions and incubation times vary with the antibody source, antigen concentration and detection system used. Hyperimmune polyvalent or monospecific sera or monoclonal ascites are usually diluted 1/100 to 1/200 for incubation times of 1–2 h, or diluted further (1/500 to 1/1000 or greater) for overnight incubation. Monoclonal antibody supernates are usually used at lower dilutions (1/5 to 1/20), with incubation for several hours or overnight. Incubation is usually at room temperature with gentle shaking; care should be taken to ensure that membranes do not dry out, since this causes background problems.

After incubation with primary antibody, blots are washed repeatedly in PBS to remove unbound antibody, and are then incubated for 1–2 h with an appropriate antibody conjugate. Conjugate dilutions of 1/2000 or greater are normally used. Conjugates may be species-specific antibody conjugates, or conjugates based on *S. aureus* protein A. Generally, specific antibody conjugates offer better sensitivity (Hames, 1990). Conjugates are tagged with a variety of molecules to allow detection of bound antibody. These include enzymes such as horseradish peroxidase or alkaline phosphatase which, in the presence of an appropriate substrate and chromogen, produce a coloured band or spot where antigen and antibody have combined on the membrane. Alkaline phosphatase conjugates are usually more sensitive than those based on horseradish peroxidase, and profiles generated with these conjugates do not fade on storage. Biotinylated and immunogold conjugates have also been used. Iodinated protein A conjugates have been used in the past, while detection systems which use chemiluminescence have been developed and marketed more recently. The latter systems are reported to enhance sensitivity significantly. Use of radiolabels introduces an additional safety hazard, and detection systems using radiolabels or chemiluminescence both require access to darkroom facilities.

After development of blots to visualize the antigenic profile, membranes are washed and dried. They are then photographed to provide

(a)

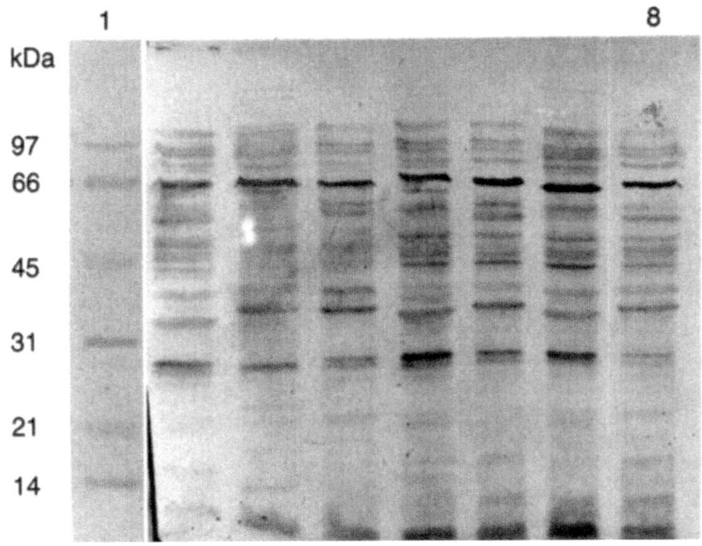

Figure 6.1 Examples of immunoblots (a) Immunoblot of *E.coli* whole-cell lysates reacted with normal human serum as a source of anti-*E.coli* antibodies. Molecular mass markers, stained with Naphthalene Blue 12B are shown in lane 1.

a permanent record. Alternatively, membranes may be re-probed with other antibodies to obtain additional information for a different set of antigens.

Typical examples of immunoblots generated using whole bacterial cells and polyclonal and monoclonal antibodies are shown in Fig 6.1.

(c) Analysis of immunoblots

Immunoblotting profiles generated with polyvalent sera are similar in general appearance to those produced on stained SDS-PAGE gels (Chapter 5). Profiles may be complex, with a diverse range of antigens recognized in each isolate. Strains are compared side-by-side on the

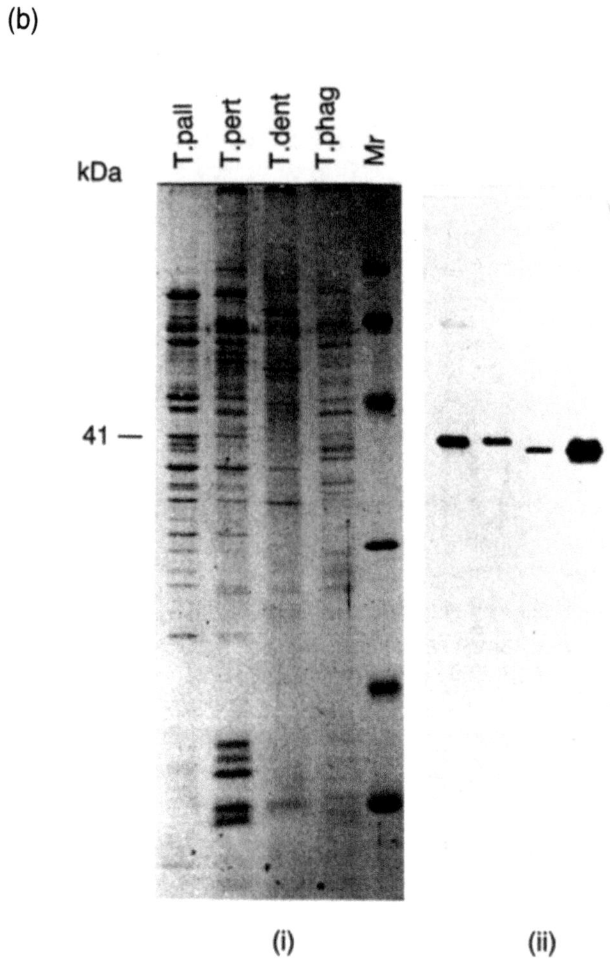

Figure 6.1 (b) SDS-PAGE protein profiles (left) and immunoblot (right) of four *Treponema* spp. isolates. The immunoblot was reacted with a monoclonal antibody produced against a 41 kDa *T. pallidum* polypeptide antigen. (*T. pall* = *T. pallidum*; *T. pert* = *T. pertenue*; *T. dent* = *T. denticola*; *T. phag* = *T. phagedenis*; Mr = molecular mass markers).

(c)

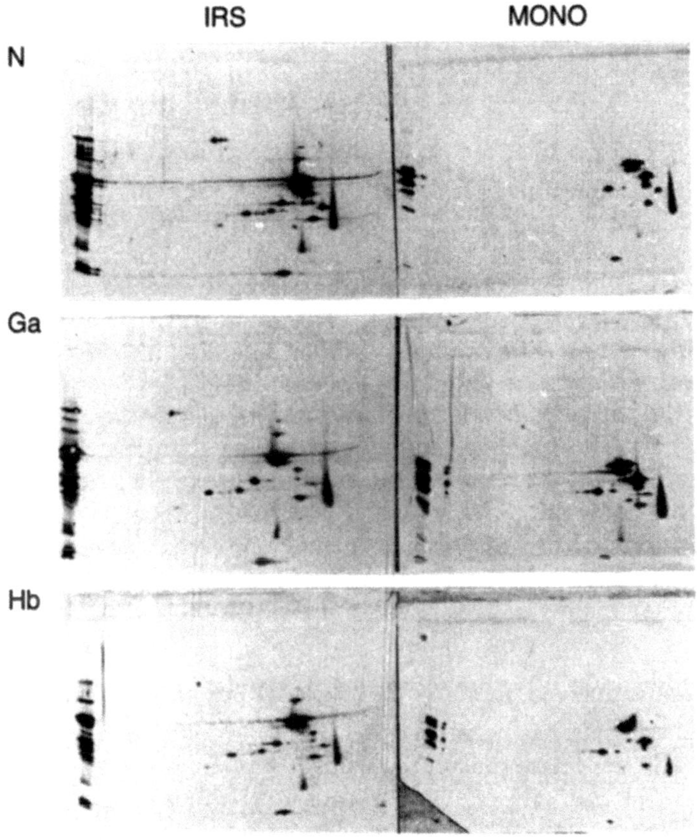

Figure 6.1 (c) Two dimensional immunoblot of a *T. pallidum* strain (N) and two *T. pertenue* strains (Ga and Hb) reacted with a polyclonal rabbit anti-*T. pallidum* serum (IRS) or a pool of anti-*T. pallidum* monoclonal antibodies (MONO).

same gel, and the pattern of common and antigenic bands is recorded by eye. To date, automated analysis of immunoblot profiles has not been generally available. Molecular masses of antigens are calculated by reference to appropriate markers electrophoresed on the same gel, though uneven shrinkage of the membranes may make accurate mass determinations difficult.

Where immunoblots are stained with monoclonals, single antigenic bands can usually be detected in each gel track, and the presence or absence of reactivity and any variation in the mass of bands produced is noted. Multiple bands may sometimes be produced on immunoblots stained with monoclonals. This may result from degradation of the antigen before electrophoresis, or from the presence of immature or unprocessed forms of the same antigen in the preparation.

The commonest technical problems associated with immunoblotting include poor transfer and distortion of profiles because of trapping of air bubbles within the blotting 'sandwich', or because of incomplete equilibration of the gel with the transfer buffer before blotting. High background levels result from inadequate blocking, or the use of low antibody or conjugate dilutions, and/or extended incubation times. In addition, polypeptide antigens of similar molecular mass may appear as single antigenic bands on immunoblots of one-dimensional SDS-PAGE gels. This problem can be overcome by changing the acrylamide concentration of the resolving gel or by the use of 2D-PAGE to separate antigens. Some microbial antigens, particularly some lipoproteins, glycoproteins and LPS molecules, may not be resolved into single antigenic bands on immunoblots. These antigens migrate to produce a smear which may make strain comparison based on these antigens difficult.

(d) Applications of immunoblotting to typing of microorganisms

Immunoblotting has been used for strain differentiation of a variety of medically important microorganisms (Burnie and Matthews, 1987). When used with polyvalent sera, produced either by immunization of laboratory animals (Burnie and Matthews, 1987), or obtained from patients infected naturally with the pathogen (Roggen *et al.*, 1992), antigenic profiles or fingerprints similar in overall appearance to those of stained SDS-PAGE gels are obtained. This approach has been used to detect reproducible differences between strains of *S. aureus* (Lee and Burnie, 1988), *H. pylori* (Burnie *et al.*, 1988) and *C. albicans* (Lee *et al.*, 1986) where SDS-PAGE protein profiles alone were too similar for reproducible strain differentiation. Similarly, immunoblotting of LPS preparations has been used to differentiate strains of Gram-negative bacteria. Perez-Perez and Blaser (1987) used proteinase K digestion and immunoblotting to investigate serotyping of *H. pylori*. These workers also found that, although strains within a particular group shared common LPS profiles, they differed in whole-cell protein fingerprints, thereby emphasizing the complementary nature of the two techniques for strain typing.

Polyclonal sera have been used in combination with immunoblotting for epidemiological investigations of disease outbreaks. Poxton *et al.* (1984), used this technique to investigate strains of *Clostridium difficile* causing antibiotic-associated colitis and diarrhoea. More recently, Kato *et al.* (1993) have extended this type of study to sub-group *C. difficile* isolates assigned previously to one of 10 different serotypes (Delmee *et al.*, 1985). However, over 40% of unrelated clinical strains investigated could not be subtyped with the available antisera, and these workers acknowledged the need to produce additional sera to improve the usefulness of this approach.

Immunoblotting has also been employed to investigate the validity of serogroups of some Gram-negative bacteria established with conventional typing sera. For example, Rimler (1990) found that LPS from isolates of *Pasteurella haemolytica* assigned originally to different serotypes showed very similar immunoblot profiles, supporting results of earlier studies which suggested that the assigned serotypes were artefactual (Rimler *et al.*, 1984).

The combination of monoclonal antibodies and immunoblotting has proven very useful for typing and epidemiological purposes. Monoclonals to surface-associated proteins were used to confirm differences in SDS-PAGE profiles for geographically diverse isolates of the bacterium *Borrelia burgdorferi* (Barbour *et al.*, 1985). Recently, Achtman *et al.* (1992) used a large panel of monoclonals to investigate both LPS and outer membrane protein variation in clones of *N. meningitidis* serogroup A causing epidemic disease in Africa, to determine the potential mechanisms of spread of the pathogen within the population.

Immunoblotting has also been used for the investigation of both antigen and epitope specificity of microbial antigens for the selection of useful diagnostic reagents. For example, the requirement for large numbers of individual antibodies raised against individual species can be avoided if genus-specific antibodies are available. Immunoblotting has been used to select genus-specific epitopes on a 60 kDa heat-shock protein of *Legionella* spp. (Steinmetz *et al.*, 1992), and conserved and distinct epitopes on viruses, including ovine lentiviruses (Marcom *et al.*, 1991). Both immunoblotting and dot-blotting have been used by Gu *et al.* (1992) to identify anti-*N. meningitidis* LOS monoclonal antibodies for typing purposes.

Other examples of organisms where immunoblotting has been used for strain differentiation and typing are listed in Table 6.2.

Table 6.2 Examples of organisms for which immunoblotting or dot-immunoblotting has been used in strain differentiation and typing

Organism	Reference
Bordetella pertussis	Peppler (1984)
Borrelia burgdorferi	Barbour *et al.* (1985)
Borrelia hermsii	Schwan *et al.* (1992)
Brucella spp.	Garin-Bastuji *et al.* (1990)
Campylobacter pylori	Burnie *et al.* (1988)
Campylobacter pyloridis	Perez-Perez and Blaser (1987)
Campylobacter spp.	Perez-Perez and Blaser (1985)
Candida albicans	Lee *et al.* (1986)
Clostridium difficile	Kato *et al.* (1993)
Coagulase-negative staphylococci	Dryden *et al.* (1992)
Echovirus type 25	Peigue-Lafeuille *et al.* (1991)
Francisella tularensis	Fulop *et al.* (1991)
Haemophilus ducreyi	Roggen *et al.* (1992)
Legionella spp.	Steinmetz *et al.* (1991)
Moraxella catarrhalis	McKenzie *et al.* (1992)
Mycoplasma arthritidis	Washburn and Hirsch (1990)
Neisseria meningitidis	Achtman *et al.* (1991) Gu *et al.* (1992)
Neorickettsia helminthoeca	Rikihisa (1991)
Ovine lentivirus	Marcom *et al.* (1991)
Pasteurella haemolytica	Lacroix *et al.* (1993)
Pasteurella multocida	Rimler (1990)
Pneumocystis carinii	Lundgren *et al.* (1992)
Salmonella spp.	Luk and Lindberg (1991)
Serratia marcescens	Hamadeh *et al.* (1990)
Staphylococcus aureus	Gaston *et al.* (1988)
Streptococcus pneumoniae	Hakenbeck *et al.* (1991)
Treponema pallidum	Cockayne *et al.* (1989) Noordhoek *et al.* (1990)
Yersinia enterocolitica	Sorey *et al.* (1990)

6.3.3. OTHER MODERN METHODS USED IN SEROTYPING

(a) Enzyme-linked immunosorbent assay (ELISA)

For some applications, immunoblotting or dot-blotting are not ideal methods for antigenic analysis of microorganisms. In particular, some epitopes on protein antigens that may be of potential use for typing may be destroyed during sample preparation for SDS-PAGE. Other antigens, such as polysaccharide antigens, are not generally amenable to analysis by immunoblotting. For some organisms, e.g. many viruses, antigen may be available in limited quantities, making immunoblotting impractical. Finally, analysis of large numbers of strains and antibodies by these techniques is cumbersome and time-consuming. ELISA techniques can be used to overcome some of these problems.

For ELISA, in contrast to conventional immunoblotting, purified antigen – or, more commonly, a mixture of antigens such as whole microbial cells or whole cell sonicate – is usually analysed in a native or non-denatured form. The antigen is coated on to a plastic surface, rather than a membrane as for immunoblotting. Modular plastic plates containing individual wells arranged in a standard 8 × 12 format are available commercially, allowing 96 individual antigen-antibody reactions to be performed on a single plate. Other formats, such as individual plastic cuvettes, have also been used (Murphy and Apicella, 1985). Once antigen is bound to the plastic solid phase, the techniques used for antibody incubation and detection are essentially similar to those described for immunoblotting. Direct ELISA uses antibodies conjugated directly to an appropriate label to allow antibody binding to be detected. More commonly, an indirect or sandwich-ELISA technique is used, where binding of unlabelled primary antibody is detected with an appropriately labelled antibody conjugate. In contrast to immunoblotting, however, the chromogen used to detect conjugate binding gives a soluble coloured product. Colour development in individual wells of the ELISA plate is assessed by eye or, more commonly, quantified with a commercially available ELISA plate reader – a modified spectrophotometer. It is necessary to determine cut-off values to discriminate between true positive reactions and background reactivity by including appropriate positive and negative controls with each series of assays. A more detailed account of general ELISA techniques is provided by Wreghit and Morgan-Capner (1990).

The development of monoclonal antibodies has enabled molecular analysis of microbial antigens and individual epitopes by ELISA techniques, and this methodology has been particularly useful with bacteria and viruses. For example, Wedege *et al.* (1991) used ELISA and monoclonals to investigate relationships between different serotypes of *N. meningitidis*. Polyclonal sera have also been used in combination with ELISA. A collection of non-typable *Haemophilus influenzae* strains was grouped into six serotypes with polyclonal sera, bacterial outer-membrane proteins and a modified ELISA technique (Murphy and Apicella, 1985). Monoclonals in combination with ELISA have been used for many viruses. For example, Coulson (1993) have recently used this method to type human isolates of rotavirus VP4.

Other examples of the use of ELISA for microbial typing are listed in Table 6.3.

Table 6.3 Examples of organisms where ELISA has been used for typing and strain differentiation

Organism	Reference
Bacteroides fragilis	Weintraub *et al.* (1985)
Bacteroides spp.	Ebersole *et al.* (1984)
Bovine RSV	Baker *et al.* (1992)
Brucella spp.	Garin-Bastuji *et al.* (1990)
Chlamydia trachomatis	Barnes *et al.* (1985)
Haemophilus influenzae	Murphy and Apicella (1985)
Human rotavirus VP4	Coulson (1993)
Mycobacterium avium	Denner *et al.* (1992)
Mycoplasma arthritidis	Washburn and Hirsch (1990)
Neisseria meningitidis	Abdillahi and Poolman (1987) Wedege *et al.* (1991) Gu *et al.* (1992)
Pasteurella haemolytica	Burrells *et al.* (1983)
Staphylococcus aureus	Nelles *et al.* (1985)

(b) Immunoprecipitation

This technique may be used with polyclonal sera, but may also be particularly useful when monoclonal antibodies that recognize non-denatured or conformational epitopes on microbial antigens are being investigated for typing purposes. The first stage of this technique involves the binding of an antibody to a specific antigen in solution. The antigen is usually solubilized by sonication and/or the use of detergents before incubation with the antibody. Alternatively, the antigen–antibody complex present, e.g. on a bacterial cell surface, may be solubilized once antibody binding has occurred. The second stage of the immunoprecipitation technique involves interaction of the solubilized antigen–antibody complex with *S. aureus* protein A, either on the surface of whole staphylococcal cells or agarose beads. Protein A–antibody–antigen complex can then be recovered by centrifugation and washing. Once washing is complete, this complex is solubilized and analysed by SDS-PAGE. It is possible to use simple staining techniques to detect precipitated antigens but, more commonly, antigens are radiolabelled before solubilization, and the profiles generated are analysed after autoradiography. This modification of the basic technique is called **radioimmunoprecipitation**. A more detailed description of this method is given by Goding (1983).

Immunoprecipitation may be useful where antigens of value for serotyping are present in small quantities relative to other antigens, since the protein A step effectively concentrates the antigen of interest. Problems may arise in interpretation of immunoprecipitation results if antigens are inadequately solubilized, since single antibodies may then precipitate antigen complexes.

This technique has been applied to a range of organisms. Tam *et al.* (1982) used radio-immunoprecipitation and monoclonal antibodies to serotype isolates of *Neisseria gonorrhoeae*. Several workers have used radio-immunoprecipitation in combination with other techniques for typing and strain differentiation of viruses. These include porcine group C rotavirus (Ojeh *et al.*, 1991) and bovine respiratory syncytial virus (Baker *et al.*, 1992). This technique has also been used for the identification of antigens of value for the diagnosis of *Leishmania* infections (Kutner *et al.*, 1991).

6.4 SUMMARY

Serotyping has been, and will continue to be, a powerful and versatile means of differentiating between microorganisms. In addition to their use for strain typing, antibodies – particularly monoclonals – also offer a rapid and definitive means of identifying organisms to the genus or species level. Consequently, monoclonals are being used increasingly in routine diagnostic microbiology laboratories.

Some of the original methods used for serotyping bacteria are still used widely today, and slide agglutination and immunofluorescence tests are technically simple, though time-consuming if large numbers of strains are to be examined. Immunoblotting allows the generation of complex antigen fingerprints for detailed molecular analysis of a wider range of microbial antigens of value for strain typing. ELISA tests can be used for large scale studies where polysaccharide or heat-labile protein antigens are important for strain differentiation. All of these techniques can be used with either polyclonal or monoclonal antibodies. The use of monoclonals specific for particular antigens or epitopes potentially allows more detailed analysis of the relatedness of organisms based on antigenic composition.

The major limitation for all these techniques is the availability of suitable antibodies. Many polyclonal and, increasingly, monoclonal antibodies used for routine typing and identification of medically important organisms are available commercially or through reference laboratories. Where antibodies to individual organisms are not available commercially, small-scale production of a limited range of polyclonal antisera can be undertaken in research laboratories. Use of sera raised against a limited number of isolates or serotypes may, however, result in a proportion of test strains being non-typable with the available sera. Sera collected from patients or animals infected naturally with specific pathogens provide an alternative source of antibodies for typing purposes. Collection of such sera may require ethical approval or licencing and, although useful, may be of variable composition, limited availability, and may require safety testing before use. Monoclonal antibodies, though technically more difficult to produce initially, may offer several advantages over conventional typing sera, particularly when used for identification purposes. The epitope specificity offered by monoclonals should also allow further refinement of existing typing schemes based on polyclonal sera, resulting in improved specificity and the validation of current serotype designations.

REFERENCES

Abdillahi, H. and Poolman, J.T. (1987) Whole-cell ELISA for typing *Neisseria meningitidis* with monoclonal antibodies. *FEMS Microbiology Letters*, **48**, 367–71.

Abe, C., Saito, H., Tamioka, H. and Fukasawa, Y. (1989) Production of monoclonal antibody specific for *Mycobacterium avium* and immunological activity of the affinity-purified antigen. *Infection and Immunity*, **57**, 1095–9.

Achtman, M., Wall, R.A., Bopp, M. *et al.* (1991) Variation in class 5 protein expression by serogroup A meningococci during a meningitis epidemic. *Journal of Infectious Diseases*, **164**, 375–82.

Achtman, M., Kusecek, B., Morelli, G. *et al.* (1992) A comparison of the variable antigens expressed by clone IV–1 and subgroup III of *Neisseria meningitidis* serogroup A. *Journal of Infectious Diseases*, **165**, 53–68.

Anderson, L.J., Hierholzer, J.C., Tsou, C. *et al.* (1985) Antigenic characterization of respiratory syncytial virus strains with monoclonal antibodies. *Journal of Infectious Diseases*, **151**, 626–33.

Baker, J.C., Wilson, E.G., McKay, G.L. *et al.* (1992) Identification of subgroups of bovine respiratory syncytial virus. *Journal of Clinical Microbiology*, **30**, 1120–6.

Barbour, A.G., Heiland, R.A. and Howe, T.R. (1985) Heterogeneity of major proteins of Lyme disease *Borreliae*: a molecular analysis of North American and European isolates. *Journal of Infectious Diseases*, **152**, 478–84.

Barnes, R.C., Wang, S.-P., Kuo, C.-C. and Stamm, W.E. (1985) Rapid immunotyping of *Chlamydia trachomatis* with monoclonal antibodies in a solid-phase immunoassay. *Journal of Clinical Microbiology*, **22**, 609–13.

Boiron, P., Stynen, D., Belkacem, G. *et al.* (1992) Monoclonal antibodies to a specific 54-kilodalton antigen of *Nocardia* spp. *Journal of Clinical Microbiology*, **30**, 1033–5.

Boppana, S.B., Smith, R.J., Stagno, S. and Britt, W.J. (1992) Evaluation of a microtiter plate fluorescent-antibody assay for rapid detection of human cytomegalovirus infection. *Journal of Clinical Microbiology*, **30**, 721–3.

Bouvet, P.J.M. and JeanJean, S. (1992) Evaluation of two colored latex kits, the Wellcolex colour Salmonella test and the Wellcolex colour Shigella test, for serological grouping of *Salmonella* and *Shigella* species. *Journal of Clinical Microbiology*, **30**, 2184–6.

Browning, G.F., Chalmers, R.M., Fitzgerald, T.A. and Snodgrass, D.R. (1992) Evidence for two serotype G3 subtypes among equine rotaviruses. *Journal of Clinical Microbiology*, **30**, 485–91.

Burnie, J. and Matthews, R.C. (1987) Immunoblot analysis: a new method for fingerprinting hospital pathogens. *Journal of Immunological Methods*, **100**, 41–6.

Burnie, J.P., Lee, W., Dent, J.C. and McNulty, C.A.M. (1988) Immunoblot fingerprinting of *Campylobacter pylori*. *Journal of Medical Microbiology*, **27**, 153–9.

Burrells, C., Evans. H.B. and Dawson, M. (1983) Antigenic relationship between the serotypes of *Pasteurella haemolytica* demonstrable by enzyme-linked immunosorbent assay (ELISA). *Veterinary Microbiology*, **8**, 187–98.

Carlin, N.I.A., Rahman, M., Sack, D.A. *et al.* (1989) Use of monoclonal antibodies to type *Shigella flexneri* in Bangladesh. *Journal of Clinical Microbiology*, **27**, 1163–6.

Cockayne, A., Strugnell, R.A., Bailey, M.J. and Penn, C.W. (1989) Comparative antigenic analysis of *Treponema pallidum* laboratory and street strains. *Journal of General Microbiology*, **135**, 2241–7.

Coulson, B.S. (1993) Typing of human rotavirus VP4 by an enzyme immunoassay using monoclonal antibodies. *Journal of Clinical Microbiology*, **31**, 1–8.

Cukor, G., Perron, D.M., Hudson, R. and Blacklow, N.R. (1984) Detection of rotavirus in human stools by using monoclonal antibody. *Journal of Clinical Microbiology*, **19**, 888–92.

Danielsson, D., Blomberg, B., Jarnerot, G. and Kosunen, T.U. (1988) Heterogeneity of *Campylobacter pylori* as demonstrated by co-agglutination testing with rabbit antibodies. *Scandinavian Journal of Gastroenterology*, **23**, 58–63.

Delmee, M., Homel, M. and Wauters, G. (1985) Serogrouping of *Clostridium difficile* strains by slide agglutination. *Journal of Clinical Microbiology*, **21**, 323–7.

De Macario, E.C. and Macario, A.J.L. (1983) Monoclonal antibodies for bacterial identification and taxonomy. *American Society for Microbiology News*, **49**, 1–7.

Denner, J.C., Tsang, A.Y., Chatterjee, D. and Brennan, P.J. (1992) Comprehensive approach to identification of serovars of *Mycobacterium avium* complex. *Journal of Clinical Microbiology*, **30**, 473–8.

Dromer, F., Gueho, E., Ronin, O. and Dupont, B. (1993) Serotyping of *Cryptococcus neoformans* by using a monoclonal antibody specific for capsular polysaccharide. *Journal of Clinical Microbiology*, **31**, 359–63.

Drouet, E.B., Denoyel, G.A., Boude, M. *et al.* (1991) Characterization of an immunoreactive species-specific 19-kilodalton outer membrane protein from *Helicobacter pylori* by using a monoclonal antibody. *Journal of Clinical Microbiology*, **29**, 1620–4.

Dryden, M.S., Talsania, H.G., Martin, S. *et al.* (1992) Evaluation of methods for typing coagulase-negative staphylococci. *Journal of Medical Microbiology*, **37**, 109–17.

Ebersole, J.L., Frey, D.E., Taubman, M.A. *et al.* (1984) Serological identification of oral *Bacteroides* spp. by enzyme-linked immunosorbent assay. *Journal of Clinical Microbiology*, **19**, 639–44.

Edwards, J.M.B., Taylor, C.E.D. and Tomlinson, A.H. (1982) Immunofluorescence techniques in diagnostic microbiology, in *Public Health Laboratory Service Monograph Series* **18**, (eds J.M.B. Edwards, C.E.D. Taylor and A.H. Tomlinson), HMSO, London.

Fulop, M.J., Webber, T., Manchee, R.J. and Kelly, D.C. (1991) Production and characterization of monoclonal antibodies directed against the lipopolysaccharide of *Francisella tularensis*. *Journal of Clinical Microbiology*, **29**, 1407–12.

Garcia, L.S., Shum, A.C. and Bruckner, D.A. (1992) Evaluation of a new monoclonal antibody combination reagent for direct fluorescence detection of *Giardia* cysts and *Cryptosporidium* oocysts in human fecal specimens. *Journal of Clinical Microbiolology*, **30**, 3255–7.

Garin-Bastuji, B., Bowden, R.A., Dubray, G. and Limet, J.N. (1990) Sodium dodecyl sulfate-polyacrylamide gel electrophoresis and immunoblotting analysis of smooth-lipopolysaccharide heterogeneity among *Brucella* biovars related to A and M specificities. *Journal of Clinical Microbiology*, **28**, 2169–74.

Gaston, M.A., Duff, P.S., Naidoo, J. *et al.* (1988) Evaluation of electrophoretic methods for typing methicillin-resisant *Staphylococcus aureus*. *Journal of Medical Microbiology*, **26**, 189–97.

Goding, J.W. (1983) *Monoclonal Antibodies: Principles and Practice*. Academic Press, London.

Gonzalez-Ruiz, A., Haque, R., Rehman, T. *et al.* (1992) A monoclonal antibody for distinction of invasive and noninvasive clinical isolates of *Entamoeba histolytica*. *Journal of Clinical Microbiology*, **30**, 2807–13.

Gu, X-X., Tsai, C. and Karpas, A.B. (1992) Production and characterization of monoclonal antibodies to type 8 lipooligosaccharide of *Neisseria meningitidis*. *Journal of Clinical Microbiology*, **30**, 2047–53.

Gustafsson, B., Rosen, A. and Holme, T. (1982) Monoclonal antibodies against *Vibrio cholerae* lipopolysaccharide. *Infection and Immunity*, **38**, 449–54.

Gustafsson, B., Lindquist, U. and Andersson, M. (1988) Production and characterization of monoclonal antibodies directed against *Bordetella pertussis* lipopolysaccharide. *Journal of Clinical Microbiology*, **26**, 188–93.

Hakenbeck, R., Briese, T., Chalkley, L. *et al.* (1991) Antigenic variation of penicillin-binding proteins from penicillin-resistant clinical strains of *Streptococcus pneumoniae*. *Journal of Infectious Diseases*, **164**, 313–9.

Hamadeh, R.M., Mandrell, R.E. and Griffiss, J.M. (1990) Immunophysical characterization of human isolates of *Serratia marcescens*. *Journal of Clinical Microbiology*, **28**, 20–6.

Hames, B.D. (1990) One-dimensional polyacrylamide gel electrophoresis, in *Gel Electrophoresis of Proteins – A Practical Approach* (eds B.D. Hames and D. Rickwood), IRL Press, Oxford, pp. 1–148.

Hamilton, A.J., Bartholomew, M.A., Figueroa, J. *et al.* (1991a) Murine monoclonal antibodies recognizing a non-capsular antigen that distinguishes between *Cryptococcus neoformans* var. *neoformans* and *C. neoformans* var. *gattii*. *Transactions of the Royal Society for Tropical Medicine and Hygiene*, **85**, 123–7.

Hamilton, A.J., Bartholomew, M.A., Fenelon, L. *et al.* (1991b) Preparation of monoclonal antibodies that differentiate between *Histoplasma capsulatum* variant *capsulatum* and *H. capsulatum* variant *duboisii*. *Transactions of the Royal Society of Tropical Medicine and Hygiene*, **85**, 425–8.

Hamilton, A.J., Harada, T., Bartholomew, M.A. *et al.* (1991c) Preparation of murine monoclonal antibodies against the yeast phase of the dimorphic fungus *Sporothrix schenckii*. *Transactions of the Royal Society of Tropical Medicine and Hygiene*, **85f**, 734–7.

Hanham, C.A., Zhao, F., Shaw, J.J. and Lainson, R. (1991) Monoclonal antibodies for the identification of New World *Leishmania*. *Transactions of the Royal Society for Tropical Medicine and Hygiene*, **85**, 220–1.

Hasenclever, H.F. (1965) The antigens of *Candida albicans*. *American Review of Respiratory Diseases* (Supplement) **92**, 150–8.

Hook, E.W., Roddy, R.E., Lukehart, S.A. *et al.* (1985) Detection of *Treponema pallidum* in lesion exudate with a pathogen-specific monoclonal antibody. *Journal of Clinical Microbiology*, **22**, 241–4.

Joly, J.R. and Winn, W.C. (1984) Correlation of *Legionella pneumophila* defined by monoclonal antibodies with epidemiological classification of cases and environmental sources. *Journal of Infectious Diseases*, **150**, 667–71.

Kaplan, D.S. and Picciolo, G.L. (1989) Characterization of instrumentation and calibrators for qualitative microfluorimetry for immunofluorescence assays. *Journal of Clinical Microbiology*, **27**, 442–7.

Karim, Q.N., Finn, G.Y., Easmon, C.S.F. *et al.* (1989) Rapid detection of *Haemophilus ducreyi* in clinical and experimental infections using monoclonal antibody: a preliminary evaluation. *Genitourinary Medicine*, **65**, 361–5.

Karimi, S.T., Schloemer, R.H. and Wilde C.E. (1989) Accumulation of chlamydial lipopolysaccharide antigen in the plasma membrane of infected cells. *Infection and Immunity*, **57**, 1780–5.

Kato, H., Cavallaro, J.J., Kato, N. *et al.* (1993) Typing of *Clostridium difficile* by western immunoblotting with 10 different antisera. *Journal of Clinical Microbiology*, **31**, 413–5.

Kaufmann, F. (1972) *Serological Diagnosis of* Salmonella *Species. Kaufmann-White Scheme*. Munksgaard International Publishers Ltd, Copenhagen.

Knapp, J.S., Tam, M.R., Nowinski, R.C. *et al.* (1984) Serological classification of *Neisseria gonorrhoeae* with use of monoclonal antibodies to gonococcal outer membrane protein 1. *Journal of Infectious Diseases*, **150**, 44–8.

Kohler, G. and Millstein, C. (1975) Continuous culture of fused cells secreting antibody of predefined specificity. *Nature*, **256**, 495–7.

Kutner, S., Pellerin, P., Breniere, S.F. *et al.* (1991) Antigen specificity of the 72-kilodalton major surface glycoprotein of *Leishmania braziliensis braziliensis*. *Journal of Clinical Microbiology*, **29**, 595–9.

Lacroix, R.P., Duncan, J.R., Jenkins, R.P. *et al.* (1993) Structural and serological specificities of *Pasteurella haemolytica* lipopolysaccharides. *Infection and Immunity*, **61**, 170–81.

Lee, W. and Burnie, J.P. (1988) Fingerprinting methicillin-resistant *Staphylococcus aureus* by the immunoblot technique. *Journal of Medical Mircrobiology*, **25**, 261–8.

Lee, W., Burnie, J.P. and Matthews, R.C. (1986) Fingerprinting *Candida albicans*. *Journal of Immunological Methods*, **93**, 177–82.

Le Minor, L. and Rhode, R. (1989) *Guidlines for the preparation of* Salmonella *antisera*. WHO Collaborating Centre for Reference and Research on Salmonella, Pasteur Institute, Paris.

Lipson, S.M., Salo, R.J. and Leonard, G.P. (1991) Evaluation of five monoclonal antibody-based kits or reagents for the identification and culture confirmation of Herpes simplex virus. *Journal of Clinical Microbiology*, **29**, 466–9.

Luk, J.M.C. and Lindberg, A.A. (1991) Anti-*Salmonella* lipopolysaccharide monoclonal antibodies: characterization of *Salmonella* BO-, CO-, DO, and EO-specific clones and their diagnostic usefulness. *Journal of Clinical Microbiology*, **29**, 2424–33.

Lund, E. and Henrichsen, J. (1978) Laboratory diagnosis, serology and epidemiology of *Streptococcus pneumoniae*, in *Methods in Microbiology*, vol. 12 (eds T. Bergan and J.R. Norris), Academic Press, London, pp. 241–62.

Lundgren, B., Kovacs, J.A., Nelson, N.N. *et al.* (1992) *Pneumocystis carinii* and specific fungi have a common epitope, identified by monoclonal antibody. *Journal of Clinical Microbiology*, **30**, 391–5.

Marcom, K.A., Pearson, L.D., Chung, C.S. *et al.* (1991) Epitope analysis of capsid and matrix proteins of North American ovine lentivirus field isolates. *Journal of Clinical Microbiology*, **29**, 1472–9.

McKenzie, H., Morgan, M.G., Jordens, J.Z. *et al.* (1992) Characterisation of hospital isolates of *Moraxella (Branhamella) catarrhalis* by SDS-PAGE of whole-cell proteins, immunoblotting and restriction-endonuclease analysis. *Journal of Medical Microbiology*, **37**, 70–6.

Moller, L.V.M., Ruijs, G.J., Heijerman, H.G.M. *et al.* (1992) *Haemophilus influenzae* is frequently detected with monoclonal antibody 8BD9 in sputum samples from patients with cystic fibrosis. *Journal of Clinical Microbiology*, **30**, 2495–7.

Murphy, T.F. and Apicella, M.A. (1985) Antigenic heterogeneity of outer membrane proteins of nontypeable *Haemophilus influenzae* is a basis for a serotyping system. *Infection and Immunity*, **50**, 15–21.

Nelles, M.J., Niswander, C.A., Karakawa, W.W. *et al.* (1985) Reactivity of type-specific monoclonal antibodies with *Staphylococcus aureus* clinical isolates and purified capsular polysaccharide. *Infection and Immunity*, **49**, 14–8.

Noordhoek, G.T., Cockayne, A., Schouls, L.M. *et al.* (1990) A new attempt to distinguish serologically the subspecies of *Treponema pallidum* causing syphilis and yaws. *Journal of Clinical Microbiology*, **28**, 1600–7.

Nowinski, R.C., Tam, M.R., Goldstein, L.C. *et al.* (1983) Monoclonal antibodies for diagnosis of infectious diseases in humans. *Science*, **219**, 637–44.

Ojeh, C.K., Jiang, B.M., Tsunemitsu, H. *et al.* (1991) Reactivity of monoclonal antibodies to the 41-kilodalton protein of porcine group C rotavirus with homologous and heterologous rotavirus serogroups in immunofluorescence tests. *Journal of Clinical Microbiology*, **29**, 2051–5.

Ørskov, F. and Ørskov, I. (1984) Serotyping of *Escherichia coli*, in *Methods in Microbiology*, vol. 14 (ed T. Bergan), Academic Press, London, pp. 43–112.

Parker, M.T. and Collier, L.H. (eds) (1990) *Topley and Wilson's Principles of Bacteriology, Virology and Immunity*, 8th edn. Edward Arnold, London.

Peigue-Lafeuille, H., Fuchs, F., Gharabaghi, F. *et al.* (1990) Impact on routine diagnosis of echovirus infections of intratypic differentiation and antigenic variation in echovirus type 25 studied by using monoclonal antibodies. *Journal of Clinical Microbiology*, **28**, 2291–6.

Peigue-Lafeuille, H., Bailley, J.L., Fuchs, F. *et al.* (1991) Heterogeneity of capsid proteins of echovirus type 25 wild type strain and prototype strain, studied by using sodium dodecyl sulfate polyacrylamide gel electrophoresis and immunoblotting. *Journal of Clinical Microbiology*, **29**, 1780–4.

Peppler, M.S. (1984) Two physically and serologically distinct lipopolysaccharide profiles in strains of *Bordetella pertussis* and their phenotype variants. *Infection and Immunity*, **43**, 224–32.

Perez-Perez, G.I. and Blaser, M.J. (1985) Lipopolysaccharide characteristics of pathogenic *Campylobacters*. *Infection and Immunity*, **47**, 353–9.

Perez-Perez, G.L. and Blaser, M.J. (1987) Conservation and diversity of *Campylobacter pyloridis* major antigens. *Infection and Immunity*, **55**, 1256–63.

Poxton, I.R., Aronsson, B., Molby, R. *et al.* (1984) Immunochemical fingerprinting of *Clostridium difficile* strains isolated from an outbreak of antibiotic-associated colitis and diarrhoea. *Journal of Medical Microbiology*, **17**, 317–24.

Prabhakar, B.S., Haspel, M.V., McClintock, P.R. and Notkins, A.L. (1982) High frequency of antigenic variants among naturally occuring human coxsackie B4 isolates identified with monoclonal antibodies. *Nature*, **300**, 374–6.

Rikihisa, Y. (1991) Cross-reacting antigens between *Neorickettsia helminthoeca* and *Ehrlichia* species, shown by immunofluorescence and western immunoblotting. *Journal of Clinical Microbiology*, **29**, 2024–9.

Rimler, R.B. (1990) Comparisons of *Pasteurella multocida* lipopolysaccharides by sodium dodecyl sulfate-polyacrylamide gel electrophoresis to determine relationship between group B and E hemorrhagic septicaemia strains and serologically related group A strains. *Journal of Clinical Microbiology*, **28**, 654–9.

Rimler, R.B., Rebers, P.A. and Phillips, M. (1984) Lipopolysaccharides of the Heddleston serotypes of *Pasteurella multocida*. *American Journal of Veterinary Research*, **45**, 759–63.

Roggen, E.L., De Breucker, S., Van Dyck, E. and Piot, P. (1992) Antigenic diversity in *Haemophilus ducreyi* as shown by western blot (immunoblot) analysis. *Infection and Immunity*, **60**, 590–5.

Schwann, T.G., Gage, K.L., Karstens, R.H. *et al.* (1992) Identification of tick-borne relapsing fever spirochete *Borrelia hermsii* by using a species-specific monoclonal antibody. *Journal of Clinical Microbiology*, **30**, 790–5.

Sorey, M.-P., Tollenaere, J., Laszlo, C. *et al.* (1990) Detection of pYV^{+} *Yersinia enterocolitica* isolates by P1 slide agglutination. *Journal of Clinical Microbiology*, **28**, 2403–8.

Steinmetz, I., Rheinheimer, C., Hubner, I. and Bitter-Saumann, D. (1991) Genus-specific epitope on the 60-kilodalton *Legionella* heat shock protein recognized by a monoclonal antibody. *Journal of Clinical Microbiology*, **29**, 346–54.

Suchland, R.J. and Stamm, W.E. (1991) Simplified microtiter cell culture method for rapid immunotyping of *Chlamydia trachomatis*. *Journal of Clinical Microbiology*, **29**, 1333–8.

Tam, M.R., Buchanan, T.M., Sandstrom, E.G. *et al.* (1982) Serological classification of *Neisseria gonorrhoeae* with monoclonal antibodies. *Infection and Immunity*, **36**, 1042–53.

Terpstra, W.J., Korver, J.H., van Leeuwen, J. *et al.* (1985) The classification of Sejroe group of *Leptospira interrogans* with monoclonal antibodies. *Zentralblatt für Bakteriologie und Hygiene*, **259**, 498–506.

Towbin, H. and Gordon, J. (1984) Immunoblotting and dot immunoblotting – current status and outlook. *Journal of Immunological Methods*, **72**, 313–40.

Towbin, H., Staehelin, T. and Gordon, J. (1979) Electrophoretic transfer of proteins from polyacrylamide gels to nitrocellulose: procedure and some applications. *Proceedings of the National Academy of Sciences of the USA*, **76**, 4350–4.

Washburn, L.R. and Hirsch, S. (1990) Comparison of four *Mycoplasma arthritidis* strains by enzyme immunoassay, metabolism inhibition, one- and two-dimensional electrophoresis and immunoblotting. *Journal of Clinical Microbiology*, **28**, 1974–81.

Wedege, E., Caugant, D.A., Froholm, L.O. and Zollinger, W.D. (1991) Characterization of serogroup A and B strains of *Neisseria meningitidis* with serotype 4 and 21 monoclonal antibodies and by multilocus enzyme electrophoresis. *Journal of Clinical Microbiology*, **29**, 1486–92.

Weintraub, A., Larsson, B.E. and Lindberg, A.A. (1985) Chemical and immunochemical analyses of *Bacteroides fragilis* lipopolysaccharides. *Infection and Immunity*, **49**, 197–201.

Wreghit, T.G. and Morgan-Capner, P. (eds) (1990) *ELISA in the Clinical Microbiology Laboratory*. Public Health Laboratory Service Board, London.

Yagi, S., Schnurr, D. and Lin, J. (1992) Spectrum of monoclonal antibodies to coxsackievirus B–3 includes type- and group-specific antibodies. *Journal of Clinical Microbiology*, **30**, 2498–501.

Final summary and future prospects

7

This book has been concerned with the application of molecular methods to both the identification and typing of microoganisms. However, in reality, typing is simply an extension of identification in that the whole microbial population can be considered as being made up of a deep clonal structure that has evolved since the origin of microbial life, overlaid with more recent and subtle changes selected by the environment and by the movement of genes from clone to clone by processes of genetic recombination. The aim of any identification and typing method is to generate 'windows' by which this clonal structure can be examined in either gross ('identification') or fine ('typing') detail. Thus, once 'identification' has been achieved, the aim of a 'typing' method is to discriminate between a collection of 'identified' strains in such a manner as to either refute or confirm the closeness of individual members. In addition, the ideal typing scheme will have an external frame of reference to aid in the interpretation of anomalies.

So far as 'identification' is concerned, the application of molecular methods has challenged some long-established taxonomic groupings based on phenotypic properties, and has resulted in the revision of certain classification schemes and the recognition of 'new' relationships. It has been recognized that nucleic acid hybridization and sequencing procedures provide the scientific 'gold standard' by which all other identification schemes should be judged. Nevertheless, it is inevitable that some existing taxonomic schemes will continue to exist, albeit based on flawed phenotypic foundations.

Although nucleic acid sequencing procedures would ideally provide the main basis for identification of microoganisms, and an automated rapid sequencing technique for 'routine' use is now a realistic goal (Chapter 4), at the present time these procedures are not suitable for normal identification purposes. However, as described in Chapter 3, hybridization tests involving specific nucleic acid probes are now used extensively by many laboratories to identify particular microoganisms in a wide variety of samples. Many kits are available commercially for performing such identification tests, the latest versions of which now combine the advantages of high specificity and speed (1–2 h) with the convenience of non-radioactive labelling and detection systems. Hybridization procedures do not distinguish between viable or non-viable organisms (a fact which may be either advantageous or disadvantageous, depending upon the precise circumstances of the test and the microorganism involved), but such kits seem likely to gain increasingly wide acceptance, particularly when aimed at the identification of microorganisms which are difficult or impossible to cultivate in the laboratory.

It has been stated that 'the goal of DNA probe technology is to eliminate routine cultures, whether they be bacterial, viral, or fungal'. While this goal will certainly be difficult to attain in the near future, in part because of the innate conservatism of many microbiologists, the rapid evolution and commercial development of these identification assays in recent years suggests that they will displace or supplant at least some of the more cumbersome or inconvenient identification procedures currently used routinely by many laboratories.

The main disadvantage of conventional hybridization assays for identification purposes is that most microorganisms are present in the samples being examined in only limited numbers unless they are first cultured or enriched. This problem has been overcome by the introduction of nucleic acid amplification procedures, primarily PCR, although other amplification methods are also being developed commercially. Such procedures are capable, potentially, of detecting a characteristic nucleic acid sequence present in a single virus particle or cell of a particular microorganism, and have revolutionized the approach to rapid microbial detection and identification in numerous laboratories throughout the world. The visualization of a specific amplification product on an agarose gel is often sufficiently diagnostic for most identification purposes, but amplification is sometimes combined with a hybridization assay to confirm the specificity of the reaction. The main disadvantages of amplification methods when used for identification purposes are the need for specific primers to direct the reaction, the detection of dead as well as viable microorganisms, the lack of quantification, and the fact that the exquisite sensitivity of these techniques can easily result in false-positive reactions unless appropriate stringent laboratory procedures are followed carefully.

Alternative approaches to identification that do not involve the study of nucleic acids may use antibodies or protein fingerprinting techniques. Antibodies specific for antigens present in particular species or genera can be used for direct identification of microorganisms. The antigenic composition of most microorganisms is usually fairly stable, though some organisms exhibit marked antigenic variation which may limit the use of antibodies for identification purposes. Usually, organisms are identified following isolation in the laboratory, whereupon antibodies are used in conjunction with a limited range of other morphological or phenotypic characters to confirm the identification. In some instances it is possible to detect and identify organisms directly in clinical or environmental specimens without the necessity for isolation of the organism, but this technique is generally applicable only where organisms are present in high numbers.

A common problem encountered in the use of polyclonal sera for identification purposes is that of cross-reactivity between structurally-similar antigens from otherwise unrelated organisms. For example, some LPS O-antigens used for serotyping and identification of pathogenic Gram-negative bacteria, e.g. *Salmonella* spp., contain epitopes that are shared across a range of Gram-negative genera. It is necessary to remove antibodies to these cross-reactive antigens to ensure the specificity of most polyclonal sera used for identification purposes. This is a laborious and sometimes inexact procedure. The increasing availability of monoclonal antibodies screened for their specificity against individual species-specific antigens should allow many of the problems associated with polyclonal sera produced by traditional protocols to be eliminated.

Although protein fingerprinting with whole-cell protein profiles has been used as a research tool in bacterial systematics, the technique has not, as yet, found widespread application in the routine identification of microorganisms. The potential of this method for identifying bacteria to the species level, by comparing profiles of test strains with those of type or wild strains by means of computer databases, has been shown in several studies. However, some of these techniques require specialist dedicated equipment and the use of radiolabels with an associated safety risk. With the increasing availability of monoclonals for species identification, and the expansion of highly sensitive and discriminatory nucleic acid techniques, such as PCR, into reference and some routine laboratories, it seems unlikely that the use of protein fingerprinting for identification purposes will increase dramatically in the near future.

While some of the molecular identification methods referred to above have resulted in new and important approaches to the rapid identification of specific microorganisms, particularly those that are difficult to cultivate by conventional procedures, it is the area of typing which has seen some of the most productive applications of molecular methods to routine microbiology. The ideal typing system should be: (i) standardized; (ii) have established precision and reproducibility; (iii) produce stable results over long time periods; (iv) have sufficient discrimination to distinguish closely related organisms; and (v) be broadly applicable and inexpensive. Conventional typing procedures relying on phenotypic properties suffer from the major disadvantages that: they are subject to environmental influences, with a single environmental change possibly affecting many unrelated properties; the components of some phenotypic tests are not mutually exclusive; and the development of a typing system based on phenotypic tests for a 'new' microbial group may take many months or even years. In contrast, the molecular techniques that form the subject of this book can be carried out in such a manner as to generate relatively reproducible and stable results, and also offer the possibility of a unified approach to microbial typing that can be applied immediately to any new epidemiological problem.

Of the methods based on nucleic acid content, the analysis of plasmid profiles and fingerprints has already been applied widely and successfully to the typing of many different Gram-positive and Gram-negative bacteria, including those for which no established typing scheme exists. Although it can sometimes be difficult to extract intact plasmid DNA for subsequent analysis, these techniques can provide information on the degree of relatedness of multiple isolates within a single day. However, it should always be remembered that the extrachromosomal nature of plasmids means that it is not always true that two isolates with the same plasmid profile or plasmid fingerprint are identical, and similarly, that divergence in the plasmid content of otherwise identical isolates can, for certain plasmids, occur extremely rapidly in the absence of selective pressure.

The problem of plasmid content variations can be overcome by the generation of fingerprints based on restriction endonuclease digestion of nucleic acid from the microbial chromosome. While it should be noted that not all restriction enzymes penetrate the clonal structure of a population of isolates equally, the main problem with chromosomal methods is the excess of information generated, with a corresponding low 'signal to noise' ratio, and resulting difficulties in interpretation. By means of suitable 'rare-cutting' restriction endonucleases, it is possible to generate a simplified fingerprint that can be analysed by PFGE; however, this procedure involves a relatively skill-intensive and time-consuming procedure for the extraction of chromosomal DNA, and is insensitive to small differences between isolates (thus a PFGE fingerprint difference is conclusive, but similarities are not).

An alternative approach is to sacrifice total representativeness and to examine particular defined regions of the chromosome in fine detail by means of hybridization techniques with nucleic acid probes. Such probes, including 'universal' probes for conserved or semiconserved regions, can generate a useful 'window' through which to view the clonal structure of a collection of isolates. However, hybridization analysis is still relatively complex and time-consuming, and is therefore likely to be used mainly as a typing tool for particular 'difficult' organisms.

Currently, the most powerful method for generating a defined 'window' on the clonal structure of a microbial population involves the use of nucleic acid amplification techniques such as PCR. By employing specific amplification primers, it is possible to rapidly select-out and amplify defined conserved or semiconserved regions of the microbial genome for direct RFLP analysis. Only a small initial amount of nucleic acid is required for the amplification procedure, and the whole amplification and analysis process can be completed within one day. Alternatively, random amplified polymorphic DNA fingerprints can be generated within 2–3 h by means of a primer chosen arbitrarily. This latter procedure does not require restriction endonuclease digestion to generate RFLPs, requires no previous knowledge of the organisms being investigated, and can be used to study any organism from which DNA can be prepared. The extent to which this system can fulfil the role of an ideal typing system remains to be established, but its promise is great.

All of the techniques used for polypeptide or LPS analysis also have the potential to provide information of value for strain typing purposes, but a greater and more useful level of discrimination can often be obtained by using a combination of these techniques sequentially or in parallel. Typing schemes based on SDS-PAGE protein profiles have been used extensively for both Gram-positive and Gram-negative bacteria and, to a more limited extent, for viruses and fungi. This technique is relatively simple, does not rely on the availability of specialized reagents, and is applicable to any organism where sufficient biomass can be obtained in a pure form. Where this is impracticable, e.g. with many viruses, specific radiolabelling of microbial proteins can be employed to allow analysis. Whole-cell analysis provides more information for detailed comparisons of strains, but fingerprints of subcellular fractions, including outer-membrane proteins or LPS, may provide sufficient information for some purposes.

In certain cases, one-dimensional polypeptide analysis may fail to differentiate between closely related organisms, and two-dimensional (2D) PAGE may be used instead. This technique offers significantly enhanced resolution of polypeptides, but is technically more difficult to perform, and cannot be used to compare large numbers of strains simultaneously. 2D-PAGE may allow differences between closely related organisms to be detected, but even this technique may fail to identify reproducible differences of value in typing.

Immunoblotting with polypeptide antigens separated by SDS-PAGE provides an additional means of typing organisms at the molecular level. Similar analyses can be undertaken with LPS by means of one-dimensional SDS-PAGE. ELISA offers an additional means of serotyping some organisms. In combination with polyclonal or monoclonal antibodies, immunoblotting techniques are extremely powerful and offer an additional level of discrimination to that obtained by means of stained polypeptide or LPS profiles. Although some existing serotyping schemes have been used for many decades, this field is evolving constantly as new organisms, antigens and antibodies are discovered. For many organisms, useful serotyping schemes are not available currently, and it therefore seems likely that the development of additional reagents and new serotyping schemes, and the refinement of existing ones, will continue for many years to come.

In contrast to many of the techniques based on the analysis of nucleic acids, data obtained by protein and LPS fingerprinting or immunoblotting tells us little about the detailed genetic relatedness of organisms. In some instances, organisms that share similar protein fingerprints or antigens may be quite dissimilar on a genetic basis. However, the ever-expanding use of multilocus enzyme electrophoresis is increasingly providing an insight into the extent of genetic diversity in microbial, and particularly bacterial, populations. This technique has proven to be extremely useful for identifying and marking bacterial clones for epidemiological studies, and has revolutionized our appreciation of the mechanisms involved in the spread of microbial pathogens during epidemics and pandemics. As this technique becomes used more widely, it seems certain that it will continue to play a major role in increasing our insight into the way that microorganisms evolve in response to changes in their environment, leading in some cases to outbreaks of disease in man and animals.

Conventional methods for typing microorganisms have filled a need for many years and, in many cases, will probably remain the methods of choice for the epidemiological investigation of many microbial species. However, in many investigations it is often unnecessary to define the precise 'type' of an organism, although it is usually important to answer questions about the relatedness or identity of individual isolates. The molecular fingerprinting or comparative techniques described in this book are particularly appropriate for studies of this type, and are especially useful for species that lack established or suitably discriminating typing schemes.

Regardless of the precise typing method employed, the generation of useful results requires that typing should only be performed with clear objectives in view. Different objectives will place different demands on a typing system and, in many cases, an intelligent choice of typing method can only be made once these objectives have been identified. Some typing problems will only be solved by a combination of more than one method, and the examples in the preceding chapters of particular difficulties associated with individual typing methods serve to illustrate the dangers of relying on a single system.

Each of the various typing methods has advantages and disadvantages when applied to a specific situation or particular microorganism. In addition to the ability to discriminate strains within a given species, the major considerations are probably the ease of performance and interpretation, coupled with the time taken to complete the procedure, and the availability of specific reagents and apparatus.

Although even the more complex molecular methods described in this book are applicable immediately to identification and typing problems in specialized reference or research laboratories, there is a need to simplify and/or automate the more complicated procedures before they will gain acceptance in routine diagnostic microbiology laboratories. Hybridization reactions for identification purposes, particularly 'in-solution' assays based on colorimetric or fluorescent labelling systems, have the obvious immediate potential to be automated, but no completely automated systems are available commercially at the present time.

This is an exciting time in the application of molecular methods to the identification and typing of microorganisms. Modern molecular techniques are rapid, reliable and relatively inexpensive. They are particularly suitable for the immediate investigation of microorganisms for which no conventional typing or identification system is available. It can be anticipated that even more sophisticated molecular methods will become available in the near future. In short, although the tried and tested conventional systems should not at the present time be abandoned entirely, the current situation with regard to molecular methods is definitely a case of only the end of the beginning.

Index

GPSR Compliance

The European Union's (EU) General Product Safety Regulation (GPSR) is a set of rules that requires consumer products to be safe and our obligations to ensure this.

If you have any concerns about our products, you can contact us on ProductSafety@springernature.com

In case Publisher is established outside the EU, the EU authorized representative is:

Springer Nature Customer Service Center GmbH
Europaplatz 3
69115 Heidelberg, Germany

Batch number: 10620141

Printed by Printforce, the Netherlands